DISCOVERING BOTANY

DISCOVERING
BOTANY
by P. Francis Hunt

Longman

Longman Group Limited
London
Associated companies, branches and
representatives throughout the world

First published in Great Britain 1979

Created, designed and produced by
Trewin Copplestone Publishing Ltd, London.

© Trewin Copplestone Publishing Limited 1979.
All rights reserved. No part of this book may be
reproduced or utilized in any form or by any means,
electronic or mechanical, including photocopying,
recording or by any information storage or retrieval
system, without permission in writing from Trewin
Copplestone Publishing Limited, London, England.

Phototypeset by SX Composing Limited, Rayleigh, Essex.

Printed in Italy by New Interlitho, Milan.

British Library Cataloguing in Publication Data
Hunt, Peter Francis
 Discovering botany. – (Discovering books; 2).
 1. Botany
 I. Title
 581   QK45.2

ISBN 0-582-39053-2

Contents

The Evolution of Plants	6
The Structure of a Plant	8
Roots	10
Stems	12
Leaves	14
Photosynthesis	16
Respiration and Transpiration	18
Plant Reproduction	20
Flowers	22
Pollination	24
Fruits and Seeds	26
Germination	28
Bulbs, Corms and Tubers	30
Plant Growth	32
Plant Behaviour	34
Classification and Names of Plants	36
Bacteria	38
Viruses and Phages	40
The Algae: Seaweeds	42
The Algae: Diatoms and Desmids	44
The Algae: Freshwater and Terrestrial	46
The Fungi: Moulds and Yeasts	48
The Fungi: Mushrooms and Toadstools	50
Lichens	52
Ferns and their Relatives	54
Mosses and Liverworts	56
Water Plants	58
Marshes, Bogs and Fens	60
Mountain and Seaside Plants	62
Cacti and Succulents	64
Sedges, Rushes and Reeds	66
Grasses, Bamboos and Palms	68
Fields, Heaths and Moors	70
Cereals	72
Plants as Medicines and Drugs	74
Trees	76
Shrubs and Climbers	78
Forests and Woodlands	80
Herbaceous Plants	82
New Plants from Old	84
City and Roadside Plants	86
Growing Plants Indoors	88
Growing Plants Outdoors	90
Extinction and Conservation	92
Glossary	94
Index and Acknowledgements	96

The World of Botany

The bright colours of marigolds and pansies in a window-box, or roses and chrysanthemums in parks and gardens, the shapes of trees in full leaf, and fruit ripening in an orchard are all part of the normal summer scene. In the autumn we notice the flowers fading and the trees clothed in red and gold. The delicate silhouettes of their now leafless limbs is one of the beauties of the winter landscape. And then in spring comes the renewal of life as the first green shoots poke through the cold soil. Understanding how and why these changes take place is just part of the fascinating study of botany.

There are millions of plants all around us that we never see. Microscopic bacteria and viruses live in the air, in the soil and in the water. Looking like tiny bats and balls when greatly magnified, these minute plants can be both useful and harmful to all living things. The diatoms and desmids that live in the seas are an important, if invisible, food for fish. Seen through an electron microscope, they are startling in their beauty. Algae in various colours and in all sizes from the single cell to the giant seaweed also populate the waters, and some of them have come ashore.

There are nations of plants spread around the world. The hot, dry deserts support a sparse population of cacti and other succulents. The marshes can be identified from a distance by their armies of reeds and rushes, and even the mountains and the arctic lands are not without plant life. In the forests the trees dominate the shrubs and herbaceous plants that grow under their broad canopy, and in the grasslands the waving stalks of cereal plants remind us how important plants are to all other forms of life.

Birds and insects, reptiles and mammals – all, and that includes us humans – are dependent on plants. They not only supply us with food, but also replenish the oxygen we breathe. They are a source of medicines and fuels, and the materials for clothes, buildings and hundreds of other items that we use everyday. The botanist seeks to understand the nature of plants as well as to protect and improve them. This book is written as a basic guide to botany for everyone who wants to know what botanists have discovered and how plants grow.

The Evolution of Plants

It is probably impossible for most of us to imagine what the Earth was like before life began. When we look about us we can see flowering plants, trees, insects and animals. Thousands of millions of years ago none of these things existed. Then soon after the world started to cool down, life began in the shallow seas. Ultraviolet rays from the Sun, together with lightning, caused water to combine with the methane gases and ammonia dissolved in it. They formed proteins and nucleic acids. No one knows whether this was a chemical reaction that occurred only once, or a series of reactions that took 100 million years. What we do know for certain is that the chemicals produced combined to form organisms – very simple forms of life that could grow and reproduce themselves. Over a period of millions of years these plant-like organisms developed the ability to take energy from the Sun, to turn it into food and to store it. The process by which they did this involved the production of oxygen. This oxygen entered the atmosphere and other organisms that could use it evolved.

At the same time some of the oxygen in the air changed into a special form called ozone and settled in the upper layers of the atmosphere. Ozone acts as a screen to too many harmful ultraviolet rays and other cosmic radiation. Protected by the ozone layers, the simple organisms continued to grow and adapt to a changing world.

We know that the gradual development, or evolution, of plants occurs by the accumulation of tiny, important changes in an organism's ability to survive. For example, all the offspring of any plant are similar, but some are always slightly stronger than others. The stronger plants live longer and produce more offspring, while the weaker plants gradually become extinct. However, we do not always know how or why these changes take place.

The earliest organisms 3000 million years ago were similar to the blue-green algae that exist today. But we know very little else about them or the later plants that lived during the next 2500 million years because they produced very few recognizable fossils. We know that life diversified greatly during that time because fossils from plants that lived only 500 million years ago show structures and life cycles much more complicated and varied than the early algae.

By about 350 million years ago giant horsetails, clubmosses and fern-like plants were widespread on the Earth's surface. We call the areas in which they lived the coal forests because their fossilized remains are the coal we mine and burn today.

Well-preserved fossils found in several parts of the world show that the first flowering plants appeared at least 150 million years ago. Today the flowering plants are by far the most abundant, and most of the early coal forest types have become extinct.

Flowerless plants, such as algae, fungi, mosses, liverworts and ferns still exist and are important in helping to maintain the balance of nature. But they have not remained unchanged since they first appeared on Earth millions of years ago. They are still evolving, and so too are the flowering plants and the animals.

Leaves from plants that lived and died millions of years ago have been preserved as fossils. These fossils are very similar to the leaves of ferns and trees growing today.

These tree trunks, opposite, in the Petrified Forest in Yellowstone National Park in Wyoming, are giant fossils. They were turned to stone millions of years ago by the deposits of mud and minerals that covered them.

The Structure of a Plant

The bodies of all plants and animals are made of cells. No matter how large a plant or an animal is, each living cell is very small – so small, in fact, that a thousand of them would fit side by side on the head of a pin! Although they are so tiny, the cells absorb and store foods. Then they use these foods to produce energy, to grow and to produce new cells. Some very small and simple plants, such as bacteria and certain green algae, are just a single cell. They are so minute that we can see them only when they are magnified hundreds of times. Most other plants are made of many different types of cells joined together, and are large enough for us to see.

All plant cells have a wall made of cellulose. The wall serves two purposes. First, it prevents the watery living material from oozing away. Second, it protects the cell by preventing other harmful organisms from getting in. The living matter inside the wall is composed of cytoplasm and a nucleus. Cytoplasm is a mass of proteins, carbohydrates, water and inorganic chemicals. It is just like a chemical factory that enables the cell to live and work. The nucleus is a control and information centre. It has coded instructions in the form of very complex chemicals that control how fast the cell will grow and determine whether it will become part of a leaf, or flower, or root. When a cell reproduces, it divides into two equal parts, and each part must have its own nucleus in order to live. When a plant reproduces, it is the coded instructions in the nucleus that make the new plant turn into a plant like its parents.

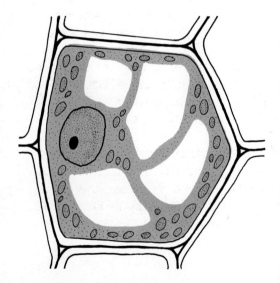

In this diagram of a plant cell the nucleus looks like a fried egg in which the 'yolk' is the mass of chromosomes. The oval shapes in the cell contain chlorophyll.

A cell from a potato plant has been cut open. The cell is dead, so there is no nucleus, but we can see the grains of stored starchy plant food.

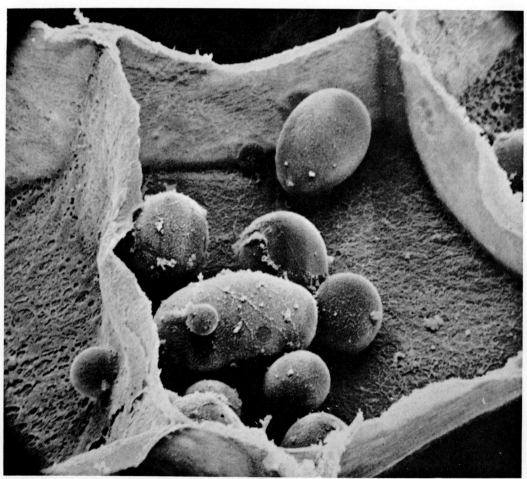

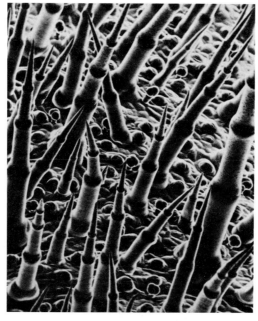

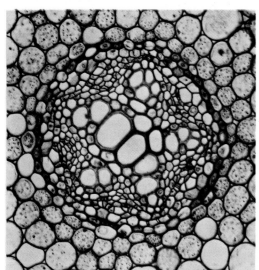

With the help of powerful, magnifying microscopes we can see parts of a plant's structure that are not visible to the unaided eye. The strange-looking objects above are really, from left to right, the tip of the female part of a flower, the surface of a leaf, and the surface of a flower petal.

Cross-sections also reveal the secrets of plant structure. Left, the veins in the centre of a root. Right, the little hairs seen in this leaf of Marram grass help to trap damp air so that the plant can live in dry soils.

A simple single-celled plant does everything it needs to do to live and grow and produce new generations inside its one cell. In more complex plants many cells are joined together to form tissues to do particular jobs. The veins in a plant, for example, are tube-shaped tissues. They are like a railway network running from the roots in the ground, up the stem, and into the leaves, where we can see them. Some veins transport the water and minerals from the roots to the cells where the plant makes its food. Others carry the food from where it is made to the rest of the plant. In certain plants the walls of the cells making up some of these veins become strengthened by layers of a chemical called lignin. The woody trunk and branches of trees and shrubs are made in this way.

Tissues can join together too. They form organs that have more complicated jobs to do. For example, the stamen of a flower is an organ that helps the plant to reproduce. However, it cannot do this job alone. It has to work with another organ. When two or more organs work together, we call them organ systems. Leaves, and flowers, and roots are all organ systems. So when we look at a single flowering plant, whether it is a tiny buttercup or a giant sunflower, we are really seeing an amazing team of organ systems built of millions of cells working together.

Roots

Tropical jungle trees are so tall that they might topple over if they did not develop supporting buttresses where the trunk meets the soil. Here we see the enormous supports of a fig tree in an African rain forest.

Roots serve two major purposes in most plants. They anchor the plant to the ground and help it to stay upright, like the foundations of a building, and provide it with water and minerals from the soil. In addition, roots may also do other jobs. In some tropical climbing plants, especially those that have very small, paper-like leaves and short stems, the roots take over the manufacture of food from the leaves, and are green. Roots are also often food storage organs, as in the case of carrots and beetroots. In tropical swamp trees the roots grow upwards and absorb oxygen. Some plants, such as ivy, have roots that help them to climb by clinging onto a suitable surface.

In a fibrous root system the base of the plant stem remains above the ground, and does not extend into the soil. It is replaced in the soil by a mass of small roots. In a tap root system, however, the base of the stem goes below the ground and becomes thickened. Grasses have fibrous roots, and parsnips, swedes and carrots are familiar plants with tap roots.

Although any single root is not always very long – it is much shorter than the plant stem – a plant usually has a great many branching roots. The total length of roots in an ordinary flowering plant can be more than 50 kilometres.

Inside all roots are the veins that carry the water and minerals up into the stem, branches and leaves. In the roots the veins are concentrated into a compact core, like the lead in the centre of a pencil. In the stem, however, the veins are arranged around the centre, like the wooden part of the pencil. Between the roots and the bottom of the stem there is a transition zone. Here the veins change from their central position in the roots to their border position in the stem.

At the tip of the roots there is an actively growing mass of cells called the root meristem. It is protected by a layer of cells that form the root cap. The cells in the meristem grow very quickly and push the capped tip further into the soil to find fresh supplies of minerals and water, and to gain greater anchorage. Further along the root, behind the cap and meristem, there is an area in which the outer cells become lengthened. They extend out into the tiny spaces between the particles of soil. They are the root hairs, the main water-absorbing parts of the roots. Root hairs have a short life. They die off as the root gets longer, and new root hairs are continually being formed.

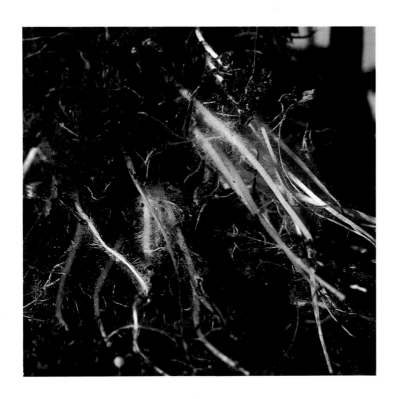

The root hairs push in between the particles of soil in their search for water and minerals for the plant.

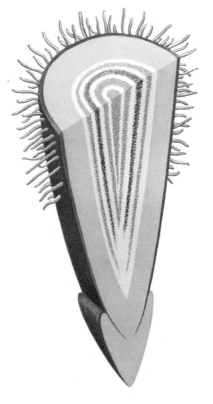

The protective cap at the base of the root advances into the soil. Behind it is the zone of actively dividing cells. In the centre the veins that will conduct water and minerals into the main body of the plant are developing. The tiny root hairs are growing on the outside of the root.

Behind the root hairs, the withered outer cells and the layer just underneath them become corky, woody or waxy. This prevents the roots from losing water and minerals back into the soil when the ground is dry. It also prevents pests from entering the roots.

In many plants the roots, and especially the root hairs, are covered with strands of fungus, like tiny cobwebs. The fungus is also found around the particles of soil. It absorbs water and minerals from the soil and passes them to the roots. Trees, heathers, orchids and many other plants have the fungus inside the cells of the roots too. In these cases both the fungus and the plant with the roots benefit. The fungus takes manufactured food from the plant, and in return passes water and minerals to the roots.

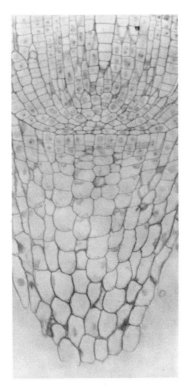

The cells of the root cap have stopped growing and are large in comparison to those immediately behind, which are growing and dividing rapidly.

Stems

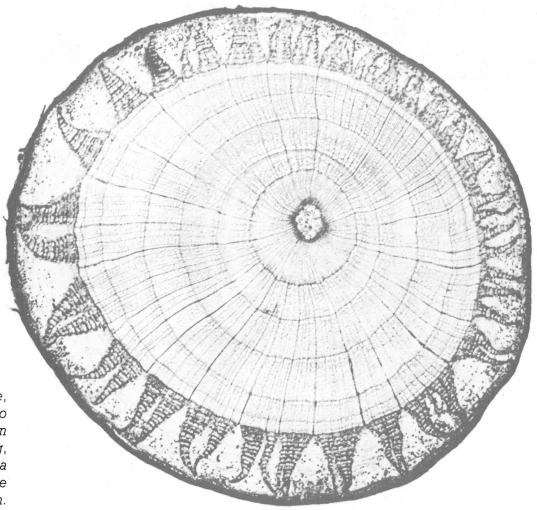

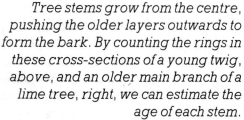

Tree stems grow from the centre, pushing the older layers outwards to form the bark. By counting the rings in these cross-sections of a young twig, above, and an older main branch of a lime tree, right, we can estimate the age of each stem.

The stems of most plants are usually standing upright, but how do they stay that way? Imagine lots of cells lying next to and on top of each other inside the stem. The walls of the cells are elastic. As the cells absorb water, they expand, and the walls stretch until they are taut, like balloons full of air. Now all the cells are pressing on each other and on the stem wall. It is the pressure that holds the stem up. When a plant does not get enough water, the cells shrink and the stem droops.

The stem supports the leaves and flowers, and can extend down into the soil to support the root system. It contains the veins that transport the minerals, water and food. Many plants have green stems, which can manufacture food and store it, especially starch and oils. In some plants, such as the cacti, the stem stores water.

Stems are usually cylindrical, but some are square or even triangular. The surface of the stem can be smooth and shiny, or hairy and ridged. Tree stems are covered with rough, woody bark, although some, like certain birches, have smooth bark.

The stem of a flowering plant bears leaves. At points called nodes the stem produces a branch, and the leaf grows at the end of it. Sometimes these branches have smaller branches of their own. In that case a leaf will grow at the end of each small branch. The dwarf shrub Butcher's Broom has a very unusual stem on which the leaves are reduced to minute scales that serve no purpose. Their place is taken by flattened, expanded, leaf-shaped green stems.

The leaves on short-lived and non-woody plants generally die and decay at the same time as the stems. However, the leaves on shrubs and trees die while the stems continue to live. When the dead leaves are shed, they leave a scar on the branch.

The 'leaves' of the Butcher's Broom are really flattened stems and bear flowers.

This young, woody stem of a horse-chestnut tree has a main bud and two secondary buds. The horseshoe-shaped scar was left by a leaf that has been shed.

At the top of each stem is a terminal growing point, or bud. Another bud grows in the angle between the leaf and the stem. This secondary bud may grow to produce a branch or a flower, or it can remain dormant. Dormant, or resting, buds either eventually wither away or grow into a branch or flower during the next season. If the terminal bud is damaged, the dormant bud can take over as the main growing point.

Many plants have climbing and scrambling stems. The stems of runner beans, for example, climb by coiling around a support, which may be another similar climbing stem! Blackberries scramble by hooking their sharp thorns onto a wall or other support, and vines and marrows have modified clinging stem tips called tendrils.

Although most stems are above ground and grow up towards the light, some, like the tubers of potatoes, grow into the soil and become swollen with stored food. Some plants have stems that grow parallel with the soil surface, and are used to reproduce the plant. Some grow on the soil surface, as in the case of strawberries, and are called runners. Others grow just below the soil surface, as in irises, and are called rhizomes.

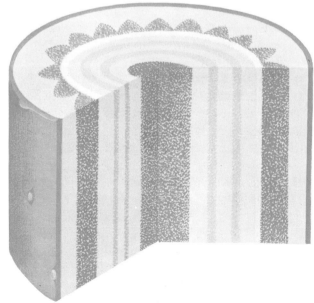

In the centre of a stem is a core of dead, empty cells. It is surrounded by a layer of thin veins conducting water and thick veins carrying food through the plant. On the outer wall, or skin, of the stem are small breathing pores.

Leaves

Leaves have two main parts: the leaf blade, which is connected to the stem of the plant by the leaf stalk. The stalk continues into the leaf blade as the main rib, and is often branched into a network of veins. Where the leaf stalk joins the stem there is a small secondary bud, and below it there is often a small leaf-like outgrowth called a stipule. In some plants the stipule is very large and completely surrounds the stems. In others it is modified to form a clinging tendril.

A leaf is constructed from layers of cells. Each layer has its own function, but leaves from the very dark and damp interiors of rain forest trees are often only two or three layers thick. The outer layer of the leaf is the epidermis, which exists like a skin mainly to hold the leaf together. The epidermis on the upper and lower surfaces of leaves generally has a waxy, waterproof outer film. On the lower surface there are small pores, called stomata, through which air and water pass. Inside the epidermis there is a layer of cells that are full of chlorophyll. This is the main food manufacturing part of the leaf. Further inside is a mass of irregularly shaped spongy cells. They can manufacture food but are mainly used to store it. The cells in these two layers are connected by veins, and between all of them there are many air spaces.

The leaves of green plants are the major site where food is manufactured. In some plants the leaves store large quantities of food, enabling the plant to survive an unfavourable season. In addition, the leaves of most ferns and related plants bear the organs necessary for the plant's reproduction. Leaves also serve as waste disposal units. Just before they are shed from a plant, leaves often change colour from green to browns, yellows and reds. This happens as their food reserves are absorbed back into the plant stem and replaced by waste materials, such as tannins. That is why in autumn in temperate regions, just before the trees lose their leaves, the forests are aglow with red, yellow and brown-leaved trees.

Leaves vary greatly in size. In some water plants they are as small as this o, while they may be as large as 2 metres in some tropical land plants. The shapes and the edges, tips and bases of leaves also vary greatly. Some of these variations seem to serve no purpose. However, they may help to prevent the plant from losing water, or they may help to expose as much of the leaf as possible to the sunlight.

When a plant absorbs too much water from the soil, the excess can be given off from the edges of the leaves. In times of drought the leaves of many plants store water and become thick and succulent. Sometimes leaves have become modified into long, thin and entwining tendrils, such as those of sweet peas. A most unusual modification is the pitcher of the various 'pitcher plants'. Here the tip of the leaf forms a container into which insects are attracted. Inside, the insects are trapped and drown in a liquid at the bottom of the 'pitcher'. Their bodies are then digested and absorbed into the plant.

Leaves come in an amazing variety of shapes. The veins usually form a network across the leaf, but in grasses and bulb plants they are parallel.

The leaves of deciduous trees are green during their life, then change colours as they die and are shed. The leaves of the water-lily and the sacred lotus float on the water.

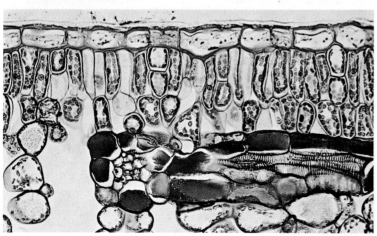

Like a single layer of bricks, the skin that prevents the loss of water covers the outer layers of cells in a leaf. Inside the leaf are the cells that contain chlorophyll and make food.

Photosynthesis

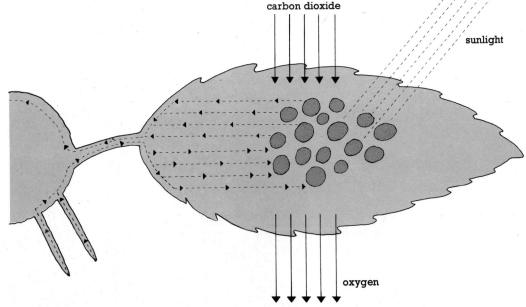

Plant cells can capture, store and change energy far more efficiently than any of man's technology. Energy from the Sun enters the cells of the green parts of green plants. It is used by the chlorophyll to change water and carbon dioxide from the air into energy-containing sugars. Oxygen is also produced and released into the air. The sugars are then passed into the stems, and carried to other parts of the plant where they are stored or used for growth, repair or movement.

Every green leaf, no matter how tiny it is, performs the miracle of photosynthesis that makes all life on Earth possible. Photosynthesis is the process by which green plants use energy from the sunlight and carbon dioxide from the air to make food. No other plants, and no animals, can manufacture food from simple substances. And even though scientists can build machines to support life in space, they cannot duplicate the food-making process of a green leaf. This means that green plants are the primary producers of all the food in the world! All other plants and all animals need green plants for their food. Even meat-eating animals get their food by eating animals that have eaten plants.

The substance that gives all green plants their colour is called chlorophyll. Molecules of chlorophyll are contained in a part of the cell called the chloroplast. When light energy falls on the green parts of a plant, the chloroplasts trap it, and the chlorophyll molecules generate minute electric currents. This electricity splits the water molecules in the plant into the oxygen and hydrogen atoms of which they are made. The oxygen is sent back into the air through the leaf's pores. The hydrogen combines with carbon dioxide, which the plant has absorbed from the air, to make sugar, or glucose.

The light energy used for photosynthesis stays locked in the sugars as chemical energy. This is stored in the plant until it is used to help the plant move, grow, or reproduce, or until the plant is eaten by an animal or dies. The bacteria, fungi, and microscopic animals in the soil feed on dead plants. Their waste products include carbon dioxide and water, which can be used again by other plants for photosynthesis. If a dead plant's remains fall into acid water or become covered by sand and mud, they may be preserved and over many thousands of years become fossils. Coal and peat are two examples of fossilized plant remains. When we burn them, we are releasing their stored energy as heat.

Because green plants use carbon dioxide and produce oxygen in photosynthesis they help to keep the balance between these two elements in the atmosphere. When people pollute the seas and cut down the tropical forests, they could be upsetting this balance. If too many green plants are destroyed, the amount of oxygen could be reduced to a level at which we may find it hard to breathe. The greater amount of carbon dioxide could also make the atmosphere overheat. If this happened, the polar ice caps might melt, raising the level of the seas and drowning half the people in the world.

Green plants make food by photosynthesis, and are the source of food for all other living things. Grazing animals eat green plant food, which their bodies use to make flesh and milk. This meat and milk, in turn, are used as food by people and other animals.

Non-green plants, like the toadstool below left, cannot make their own food, and take it from dead plants in the soil. Some plants, like the ivy below centre, make food in their green parts and store it in the non-green patches. The algae below is shown in the process of photosynthesis. As it makes its food, it releases bubbles of oxygen.

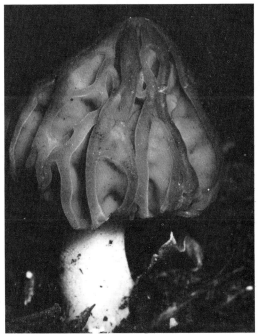

Respiration and Transpiration

Respiration is the process by which the plant absorbs oxygen and releases the energy locked in the sugars by photosynthesis. It takes place in all plants and in all animals. Burning fossilized plants, such as peat and coal, is a very rapid respiration process in which large quantities of oxygen are absorbed and a lot of energy is given off. We can think of respiration as the burning of the plant's sugars.

It happens in the cytoplasm of the cell. Each molecule of glucose is broken down with salts in the cell to make pyruric acid and other chemicals needed for plant growth and behaviour. Some energy is released at this stage, and even more is produced as the acid undergoes further chemical changes until it finally combines with oxygen to produce water. Energy can also be produced by a similar breakdown of fats and even proteins in starving plants and animals.

Transpiration is necessary not only to provide water for the plant's food-making process, but also to keep the stems and leaves rigid. The water taken from the soil by the roots passes into the veins up the stem, through the branches, and into the leaves. The pressure needed to raise water to the top leaves in the tallest trees – as much as 100 metres above the ground – is tremendous. The best man-made vacuum pump can raise water only 10 metres. But a plant can raise water to this height because of the enormous suction created by the loss of water through the leaves.

The water loss from the leaves is controlled by the action of guard cells on either side of the pores. When the Sun is shining or when the air is damp, the guard cells absorb water from other cells in the leaf. This changes their shape and forces the pores to open to allow water vapour to escape. When it is dark or the air becomes dry, the guard cells change shape again and close the pores. Many plants also have an inbuilt biological clock that opens and closes the pores regardless of the temperature, light, or dampness of the air.

When the cells in the leaves lose water, they replace it with water from the veins. The veins draw up more water through the stem from the roots, and the roots in turn take more water from the soil.

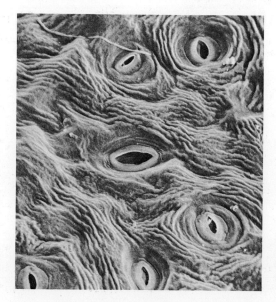
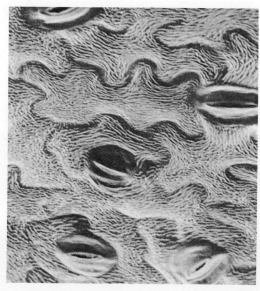

The parsley leaf, on the left, and the carrot leaf, on the right, have been magnified hundreds of times so that we can see the pores, or stomata, in both the open and closed positions.

Water enters the root hairs in the soil and travels through the veins in the roots and stem until it reaches the leaves. When the pores on the leaves are open, the water escapes as vapour into the air. This loss of water creates a suction, which enables the roots to obtain water in the first place.

Plant Reproduction

Plants can reproduce either sexually or asexually. Some plants reproduce in both ways, but others can reproduce by only one of the two methods. Sexual reproduction usually requires male and female parts of a plant, or separate male and female plants, which produce sex cells, called gametes. The gametes must come together and fuse and then grow in order to produce offspring. The simplest form of sexual reproduction can be seen in certain, thread-like algae. Identical algae line up next to each other and where they touch, their cell walls break down and the contents of one cell enters the other to produce offspring. Other algae, such as seaweeds, develop special chambers that release male and female cells into the sea.

Mosses and liverworts produce spores, which grow and develop into miniature and distinct plants. These, in turn, produce male and female cells that eventually meet and fuse to make a new spore-producing plant. This process, involving a plant that produces spores and a plant that produces sex cells, is called alternation of generations.

Ferns reproduce in a very similar way. Capsules containing spores are borne in groups on the underside of their leaves. When shed, the spores germinate into minute, flat green plants. These plants produce sex cells, which unite to form a new spore-bearing fern.

Reproduction by spores can be a very wasteful and uncertain method. A spore can retain its ability to grow for up to forty years after it has been released from the plant, but it can only survive when it starts to grow if certain conditions are favourable because it has no food reserves.

About 350 million years ago the first plants bearing seeds rather than spores appeared on Earth. These early plants were the seed-ferns. Although they existed for over 50 million years, they eventually became extinct. Their place was taken by the naked seed plants. Then about 150 million years later the closed seed or flowering plants evolved.

Most plants reproduce sexually. However, plants that live in extreme and harsh conditions, such as mountain tops, often reproduce asexually. In such environments sexual methods, which often depend on insects and birds to bring the sex cells together, are unreliable. Asexual reproduction involves the formation of a plantlet. A part of a plant is detached from the parent to form a new plant by producing roots and shoots.

One way to reproduce, or propagate, new plants is to take cuttings, which are shoots cut from the parent plant. These are placed in boxes of sandy soil to stimulate root growth. Another way is to place a leaf from a plant on the soil and make little cuts in the larger veins. After a few

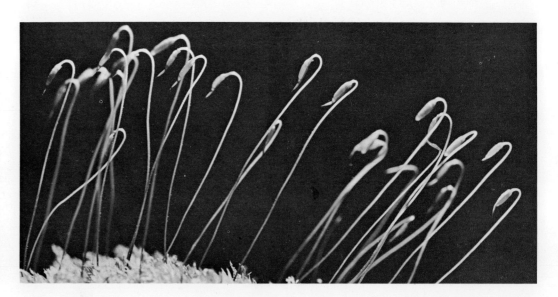

Mosses reproduce by releasing spores from long-stemmed capsules. The spores germinate to produce sex cells, which, in turn, unite to produce new moss plants.

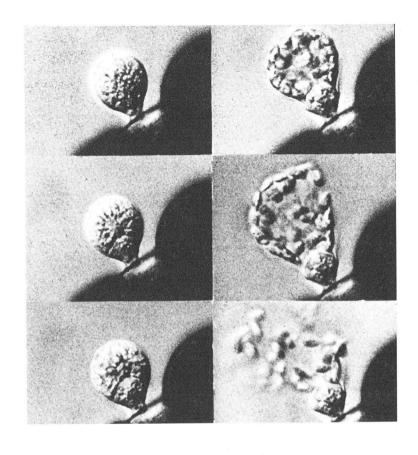

When the spore capsules on a fern leaf are ripe, they split open and cast the spores into the air.

days roots will form at these cuts. They will be followed by shoots that produce new plantlets. These are the two asexual methods usually used by gardeners.

In the 1960s a revolutionary new reproductive technique was invented. Today it is possible to produce millions of identical plants in a few weeks from one parent plant. All that is needed is the growing point, or meristem, from the stem, root or leaf. In the controlled conditions of a laboratory this mass of cells can be grown in a liquid containing carefully selected chemicals. Instead of the cells developing to form a new plant, they multiply rapidly to produce vast numbers of similar growing points. These daughter meristems either detach themselves from their parent or are cut off by a technician using a sterilized knife. Each meristem is then placed on a special jelly containing different nourishing chemicals. The chemicals now make each meristem grow into a minute plantlet, which can be removed and grown into a full-sized plant. This technique is called meristem culture and it is used mainly for orchids and apple trees.

One leaf cut into small squares, each with a major vein, and placed on damp soil will produce many new plantlets.

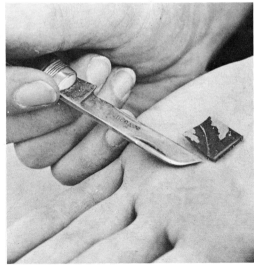

Flowers

Some flowers, like the orchids, are so beautiful that we stop simply to stare at them in wonder. Others, like the roses, have such a delightful aroma that we cannot resist poking our noses into them to inhale it more deeply. Still others, such as the dandelion, are so plentiful that we hardly notice them at all. We use flowers in many ways – to decorate a dinner table or as a gift to cheer up a sick friend. In some countries people greet their visitors with garlands of flowers, and, of course, flowers are used in making perfumes. In fact, we use flowers in so many ways that we sometimes forget that they have a purpose of their own.

The flower is the part of the plant that is responsible for reproduction. It produces the male and female cells, which unite to form the beginning of a new plant inside a seed. Each part of the flower has a special job to do.

The stamen is a long stalk that bears the male cells – the pollen – at its tip. The female cells are contained at the base of another stalk. This stalk has a sticky tip, called a stigma, to catch the pollen. The male and female stalks are usually found together on a flower. In some species, however, they grow on separate flowers on the same plant, and in a few species they grow on separate plants.

The young, unopened flower is protected by a ring of special leaves, the sepals. As the flower opens, the petals – which are really another kind of leaf – are revealed. Their job is to protect the male and female stalks, and to attract insects and animals to help in pollination. The petals and sepals can be separate or joined. They can be very different from each other or exactly the same. Usually, all the petals of a flower are similar, but in orchids, for example, the middle petal is quite different.

Some plants have single flowers on a stem. Many others bear their flowers in a group, called an inflorescence. The main purpose of these groups seems to be to attract more insects than a single flower. The colour, shape, size, number and arrangement of the parts of the flower help us to identify the many thousands of different species.

The green sepals that once covered the flower bud have been pushed back by the opening petals. Protected by the petals are the long, thin stamens and the central female stalk.

Conifers bear cones instead of flowers. The pine cone on the left is ripe and ready to shed its seeds. The cross-section on the right shows the position of the seeds on the top of the scales, and, far right, a seed.

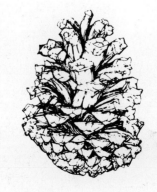 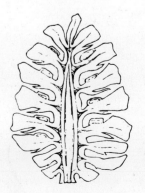

The dramatic x-ray of roses, left, reveals the structure of the flowers. A single rose, centre, grows on each branch, while the geranium, right, is an example of an inflorescence.

The flowers of short-lived plants, such as annuals, are often brilliantly coloured to attract pollinating insects to ensure the continued life of the species.

Pollination

In order for a flower to begin to produce a new plant, the pollen from the male stalk somehow must be transferred to the sticky stigma of the female stalk. This movement of the pollen is called pollination. It is followed by fertilization, which is when the pollen and the female cells unite to form a seed. Fertilization can take place only if the pollen and the female cells are from the same, or a related, species. The female stalk also has to be at just the right stage of growth for the pollen to burrow through it to find the female cells. As soon as the pollen has succeeded in its search, the seeds begin to develop and the female stalk begins to wither.

The tiny grains of pollen can be carried to the stigma by the wind, water, or an animal. The chances of the right grain of pollen being blown through the air and falling on the right stigma at the right time are not very great! Therefore, plants that are pollinated by wind and water usually produce a great deal of pollen to make sure that at least some of it will find its target. Birch, poplar and hazel trees are wind pollinated plants. Their many dangling

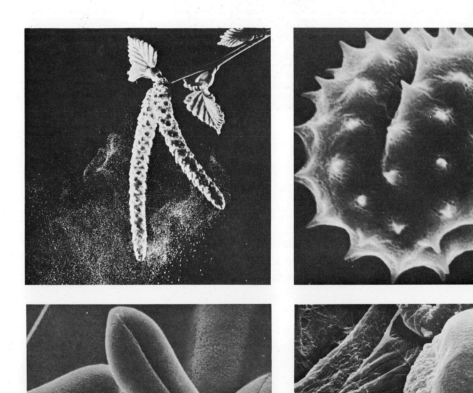

Pollen grains look just like specks of dust when they are shed from a catkin. However, when they are magnified, the pollen grains of different plants show distinct features, which can remain even when the living inside of the grain has decayed.

Like many other insects, this Cardinal Beetle has visited a flower looking for nectar. It leaves covered with pollen.

Bees return to their hive, where they store the honey they make from the nectar they have collected from flowers.

catkins shed their pollen as they are shaken by the wind. Grasses are wind pollinated too. It is the large amounts of pollen that they produce that cause hay fever, an allergic reaction that makes many people sneeze and suffer from a runny nose and sore, itchy eyes.

Many water plants shed their pollen underwater. It is carried by the water currents until it is caught by long, trailing stigmas.

Plants that depend on animals to help them pollinate do not have to produce as much pollen as those that depend on the wind or the water. These plants have developed many features to attract the animals and insects. The colour and the pattern of markings on the petals are just two of these attractions. Another is the food the animal gets as its reward for helping pollinate the plant. This can be nectar – a sweet, sugary liquid made inside the flower – or the pollen itself. As the insects gather their food, some of the pollen sticks to their bodies. It is brushed off on the stigmas as they move from flower to flower looking for more food.

The scents of flowers are also important in attracting animals. These can be very pleasant and attract butterflies and moths, or they can smell like rotting animal flesh, in which case it is blowflies and bluebottles that come.

In some plants the insect-like shape of the flower attracts male insects. They try unsuccessfully to mate with the flower, and in the process become covered with pollen. Other flowers have long, waving tassels, and many have a combination of all these features. Insects are the most common pollinators, but spiders, slugs and snails, frogs, bats and hummingbirds also help pollinate certain plants.

If pollination does not occur, the flower usually dies. Once pollen is shed, it normally has only a short life. If it does not land on a stigma, it too soon dies. However, under carefully controlled conditions in laboratories it is possible to store pollen. It can be used later for plant breeding experiments and for scientific study.

Fruits and Seeds

As soon as the pollen and the female cells have united, the seeds begin to form. In most plants the seed consists of an undeveloped, or embryo, plant with one or two tiny seed-leaves and minute roots. It has food stored either in the seed-leaves or in the tissue surrounding the embryo. The embryo and tissue are protected by a coat called the testa. There is usually a small hole in the testa to let water into the embryo when it starts to grow.

Orchid seeds have only a very tiny embryo and no supply of stored food. To make sure that the species has a chance to survive a single orchid plant will produce millions of seeds in one season. These seeds are so small that 60 million of them weigh less than a postcard. Seeds from other plants are larger, and those of the coconut tree can be as large and as heavy as a bowling ball.

Some seeds have only the testa to protect them until they begin to grow, but most are enclosed in a fruit. The fruit can be dry, like ripe pea pods and bean pods, or fleshy, like apples, pears, plums, tomatoes, oranges, lemons and grapefruit. These are all important foods for people. They develop when the cells in the female stalk holding the seeds take in moisture and grow. The cells produce sugars and chemicals that give the fruits the flavours and colours that attract the animals that eat them, and in this way help to disperse the seeds.

Seeds must be scattered as far away as possible from the parent plant so that they do not have to compete with it for water, minerals in the soil and space to grow. Plants have developed many devices for making sure that this happens. Some seeds and fruits, like those of the chestnut tree, have hooks that catch on the fur or feathers of animals, or on people's clothes. They are carried wherever the animal goes until they fall or are brushed off. In some cases seeds are carried great distances on the mud on a bird's foot or a person's shoe, or even among the dust and dirt in trucks, ships and aircraft.

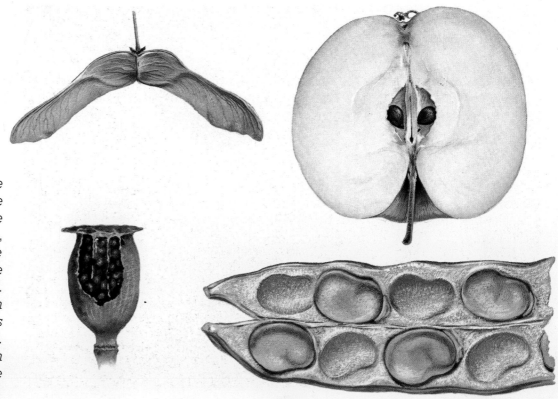

When seeds are ripe, they are scattered by the parent plant in the hope that they will find a suitable place to grow. The winged seed of a maple, top left, travels a long distance, while those of the poppy, bottom left, are shaken out of their 'pepper pot'. Apple seeds, top right, are shed when the fleshy fruit around them gets eaten and the core is thrown away. The seeds of the broad bean, bottom right, are often eaten before they are have a chance to be scattered.

Seeds can also be transported inside an animal that has eaten the fleshy fruit, such as grapes or plums or pears, surrounding them. The seeds are released when the animal excretes its waste material.

The seeds of water plants, such as the floating water lilies, contain light corky or oily substances so that they too can float and be scattered by the water.

Some seeds are transported by the wind. Very small ones are simply blown around by the air currents. Larger ones, like those of the daisy, have feathery parachutes or, like those of the sycamore, wings to help them scatter in the wind. Some plants shed their seeds when the wind shakes the 'pepper pot' capsule that holds them, as in the case of the poppy. Others explode! When the part of the fruit carrying the seeds dries, the pressure inside becomes so great that the seeds are shot into the air as if by a catapult.

Many edible fruits contain their seeds inside the juicy flesh, but strawberries have their seeds on the outside. In contrast, the horse-chestnut seed is protected by a tough, prickly coat.

A gust of wind blows the seeds off the dandelion 'clock'. With the aid of their feathery parachute tops, the seeds travel great distances.

Germination

When the air, the temperature, the amount of moisture and the light are suitable, the embryo plant in the seed can continue to grow to become a mature plant. This early stage of growth is known as germination. It can begin immediately after a seed or spore has been shed by the parent plant, but most seeds and some spores can stay dormant for a long time.

A dormant seed usually has very little water stored in it and its rate of respiration is low. The first step in germination is for the seed to absorb water and oxygen so that respiration can be speeded up. The cells in the embryo then start to expand, and soon afterwards they begin to divide. They use the food stored in the seed as their source of energy. As soon as the embryo has grown too large and the seed has taken in a lot of water, the testa splits and the young root and shoot pop out.

The root cells continue to divide and enlarge. Some of them become root hairs so that the young plant can take water and mineral salts from the soil. The seed-leaves can remain underground, but in many plants they are pushed up on the growing shoot. They turn green as chlorophyll develops in them, and then they begin to make food by photosynthesis. Eventually, the food stored in the seed-leaves is used up and they wither away. By this time the first true leaves have grown, and they take over the manufacture of food for the plant.

The growing points on the shoot and the root continue the growth of the plant above and below the ground. Before long all traces of the seed have disappeared.

Seeds can be used as food when they are dormant, as in the case of maize, wheat or rice, or when they are germinating, as with bean sprouts, mustard and salad cress. Seed germination is also a very important part of the brewing industry. Barley seeds, for example, can be harvested when they are ripe, and then they can be germinated. The starch foods that are stored in the grains are converted into sugars – malt in this case – by the chemical processes that accompany germination. When all the food has been converted in this way, the young seedlings are killed and dried. The malt is then fermented by yeast to make alcohol and carbon dioxide, which you can see in beer in the form of bubbles.

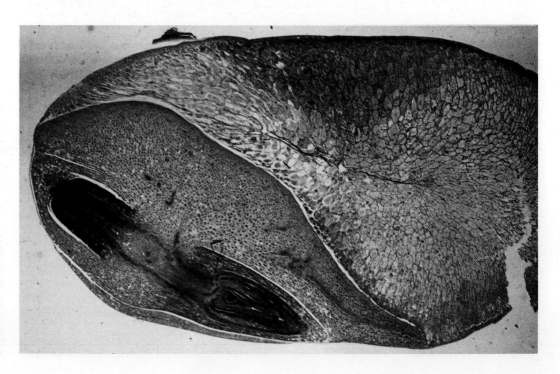

In this cross-section of a seed of maize that is just beginning to germinate the large cells of stored food look like scales on the top layer. The developing root is the rounded mass at the left, and the developing shoot is the pointed mass on the right.

A number of Acacia seeds have started to germinate. In some the seed leaves are just emerging, while in others the stems have started to stand up. In the centre is one plantlet with an upright stem and two seed leaves clearly visible. Next to it is an older plantlet with a strong, straight stem, which has produced its first true leaves.

Seeds vary greatly in size and in the conditions under which they will begin to grow. Dozens of cress seeds, bottom left, will germinate in a small glass dish, but that would not suit the coconut, centre, which has begun to grow on a tropical beach. Some seeds don't even wait to leave the parent plant, as in the case of this tomato.

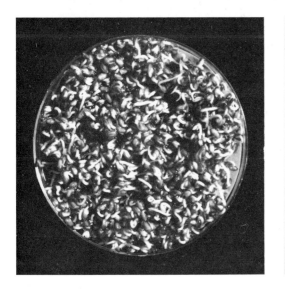

Bulbs, Corms and Tubers

Tubers of potato plants are a major source of food in many parts of the world.

Many different plants have developed the means by which they can survive underground during an unfavourable season, such as a cold, dry winter, or a hot, dry summer. These include bulbs, corms and tubers, which are organs that store food and water, and also reproduce the plant.

Bulbs are almost round storage and reproductive organs that are formed at the end of the growing season. They stay in the resting state until good growing conditions re-occur. The stems of bulbous plants, which include onions and shallots, daffodils and hyacinths, become long only at flowering time. They die away quickly as soon as the plant has finished flowering and fruiting. While this is happening one of the buds at the base of the plant grows into a very short stem. At the end of the growing season the bases of the leaves surrounding it become swollen with food reserves. The plant dies except for the short stem and the swollen, fleshy leaves.

When certain plants, such as gladiolus and crocus, have flowered, the bases of their stems become swollen with food reserves to form corms. The leaves around these stems die, leaving papery scales on the outside of the corm. Unlike bulbs, corms have no food stored in swollen leaves, but only in the swollen stem. When the conditions are suitable for the plant to start to grow again, a bud at the base of the previous year's dead main stem starts to grow and eventually produces leaves and flowers.

Tubers are either swollen roots, as in the case of the potato, or swollen stems, as is the dahlia. Some tuberous plants store food in their stems or roots only at the end of the growing season, but others always use these organs for storage. Buds from which shoots can grow appear on the outside of tubers. Usually, when the tuber is broken into smaller pieces, any piece with a bud will develop into a new plant.

People have used bulbs, corms, and tubers, such as yams, potatoes, Jerusalem artichokes and onions, for their food supply for thousands of years. They can be eaten immediately or stored for use in the future. The foods stored in these organs are usually starch, sugar or similar carbohydrates. They are often accompanied by other chemicals, many of which we use as drugs or flavourings, such as ginger and liquorice. Because these reproductive organs can be stored, they can be transported great distances and planted far from their countries of origin.

Decorative plants, such as lilies, tulips, irises and gladioli, are also examples of bulbs, corms and tubers. Today large areas in many countries are devoted entirely to producing them for sale around the world. New varieties are bred each year. Some of them can be sold for a great deal of money when they are introduced to the general market. This is nothing new however. When tulips were introduced into Europe from Turkey in the sixteenth century, they became so popular that a great craze called 'Tulipomania' swept across Holland and neighbouring countries.

Tulips are one of the many kinds of flowering plants that are grown from bulbs.

Crocuses are beautifully coloured flowers produced by corms in late winter or early spring.

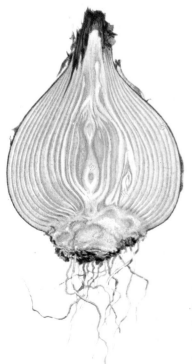

A cross-section diagram of a bulb shows the condensed stem at the base, the main bud inside, and the fleshy leaves surrounding them.

The tubers of some plants may be oddly-shaped and resemble animals or parts of the human body. This has led some people to believe that these plants, their tubers and extracts from them could be used as medicines to cure diseases that affect the part of the body that they seemed to resemble. Today, some people still dig up Bryony roots for their supposedly magical properties.

Plant Growth

When you notice that a plant is larger and perhaps heavier than it was before, you probably think to yourself that it has grown. Of course such an increase in size and weight is not always due to growth – it might only be the result of the plant absorbing a lot of water. It is rather like a man who has just had an enormous meal; he might weigh more immediately afterwards, but he hasn't really grown. Plant growth takes place when the amount of protoplasm in the plant increases. This involves the division, enlargement and development of the cells. We can see it most clearly when the plant produces a new leaf or bud.

As soon as a seed germinates, the cells of the root and the shoot start to grow. Each cell at the tip of the newly emerged young root and shoot divides very rapidly to produce more cells. The nucleus of the cell about to divide changes its chromosomes from thin interwoven threads into shorter, sausage shapes that are made of a pair of spirally coiled thicker threads. The sausage shapes arrange themselves across the centre of the nucleus. Then the pairs of threads separate and travel to either end of it. The middle of the nucleus thickens and turns into a cell wall. In this way the original cell divides into two equal parts, each with its own set of chromosomes in a nucleus and with a cell wall all round.

The newly formed cells grow to about the size of the original parent. In a quickly growing root or shoot they divide again very rapidly. Eventually, they stop dividing and begin to change to make tissues, organs, and organ systems. They do this by the addition of chemicals to the cells, which cause a change in the shape of the cell, a thickening of the cell wall, and the formation of chloroplasts and other new cell parts. Botanists are now studying why the cells of any plant alter to produce the different features of each species. Although we know how a plant develops from a seed to a mature specimen, how it does this always to the correct pattern is still a mystery.

Plants continue to grow throughout their life. The chemical DNA in the chromosomes controls the growth of a plant by manufacturing a series of plant chemicals, or hormones, in a particular order. These hormones start, stop, speed up and slow down growth. The plant also produces other chemicals that affect its growth in response to conditions such as heat and light.

The growth process does not repeat itself in plants that live only for a short time. Annuals, such as marigolds and asters, live for just one growing season, while biennials, like wallflowers and Sweet Williams, start growing one year and complete their life cycle the next. Both these kinds of plants die when they have shed their seeds. However, perennials, which include trees and shrubs, peonies and carnations for example, grow through more than two seasons. For them growth is a continually repeating process of leaf unfolding, flowering, seed ripening and shedding, and leaf-fall.

Growth is slower at night than in the day, and also during cooler periods. Plants from the cooler parts of the world usually have a dormant, or resting, phase between periods of active growth. Annuals rest when they are in the form of seeds, and perennials rest during their leafless period. Growth is continuous in tropical plants, and the rate of growth can be very high. Most plants grow less than one centimetre a day, so we hardly notice it. However,

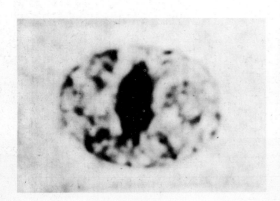
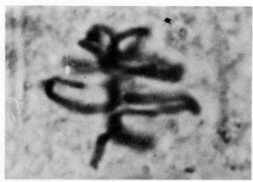
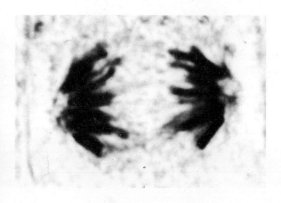

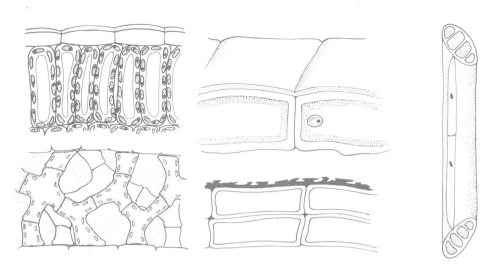
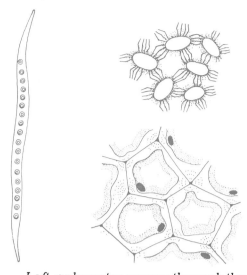

Cells serve different purposes. Top, these contain chloroplasts to trap light energy. Above, a lacework allows gases to be exchanged.

Top, these cells are covered with a waterproof layer and, above, these form bark. Right, these strengthened vein cells carry food and water.

Left, only water passes through the pores of this vein cell. Top, fibres link and strengthen cells. Above, water and sap keep stems rigid.

asparagus grown in the controlled conditions in a greenhouse have grown as much as 30 centimetres in one day. Tropical bamboos can grow more than one metre in a single day.

The life of a perennial plant, measured from the time its seeds germinate until it dies, varies greatly. In the wild the wear and tear on plants by pests, diseases and animals can cause death at any time. Lightning, gales, floods and long droughts are other natural disasters that affect many plants. There are a few trees that will grow for fifty years or even more without flowering. When they finally do flower, the whole plant dies shortly afterwards. Most trees can live for hundreds of years, and plants that reproduce themselves by runners, like strawberries, can live indefinitely.

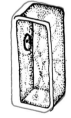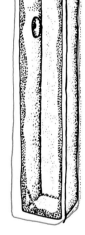

The amount of chlorophyll in a cell remains the same even when the cell grows larger.

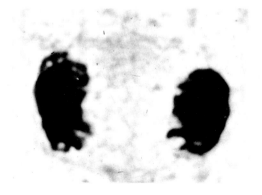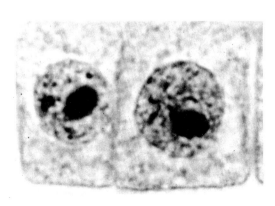

A cell can reproduce itself in about one hour. The chromosomes are clearly visible in a cell about to divide. They arrange themselves across the centre of the cell, then divide in two and move towards the ends of the cell. They begin to contract, and finally a wall is built across the centre of the old cell to form two new ones.

Plant Behaviour

Like people and other animals, plants react throughout their lives to their environment. The conditions in which any plant lives include temperature, water supply, light, air, and contact with other plants of the same species as well as with plants of other species and animals. A plant can live and grow from the time of germination through body growth, to flowering, pollination and the scattering of seeds within a certain range of these conditions. However, if they become extreme – if, for example, it is too hot or too dry or too dark at a particular time – the plant will stop its ordinary pattern of behaviour and can even die.

Any plant usually has a range of conditions under which it will grow normally, and a wider range under which it will survive but not reproduce. The conditions it needs differ at different stages in its life cycle. The germination of seeds, for example, often requires little, if any, light, but a fairly high temperature. In contrast, the growth of the seedling needs plenty of light but a lower temperature. The process of flower opening needs a higher temperature and plenty of light, while the ripening and scattering of seeds usually needs even higher temperatures. Each plant can grow successfully and produce offspring only if it has the right conditions in the right order and for the right length of time.

In extreme conditions a plant can take action to prevent damage and death, and it can even repair the damage that does occur. For example, there is a very thin film of pure water on the inner tissues of the leaves. When the air temperature falls to freezing point, this water freezes. It is instantly replaced by water taken from inside the cell. This water then freezes and is replaced in the same way. Eventually, the cell suffers from the loss of water because this increases the concentration of salt and, in turn, causes its proteins to solidify. If the water continues to be withdrawn from the cells, all the proteins will solidify and the cell will die. In order to survive many plants that live in cold areas develop a special type of sugar in their cells that holds water very tightly, and does not release it.

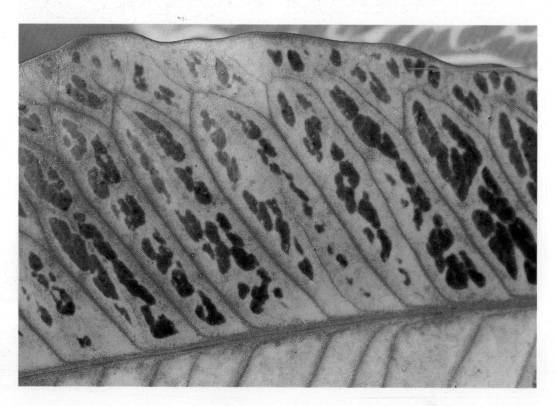

Red colouring matter develops in many leaves that are exposed to intense sunlight, such as this Croton. It protects the cells from excessive and damaging radiation.

The leaflets of the Sensitive Plant, Mimosa pudica, *will fold up when they are touched. They will open out again in about 15 minutes.*

The leaves of the Spotted Medick plant respond to light. The leaves on the left are in their daytime position, and those on the right are seen at night.

Cress seedlings grow towards the light. Those in the dish on the left have been given light in all directions. Those on the right have been placed on a window sill in a dark room and have leant over towards the light.

Plants in very hot, sunny climates develop a red colouring that prevents their rapidly growing parts, such as leaves, from getting sunburnt. They also produce a waxy covering on the leaves that retains water in extremely dry conditions.

When plants are wounded, they make a hormone called traumatin. This encourages the cells surrounding the wound to grow quickly to close the wound, and then to produce cells similar to those that were damaged. It is similar to the way our bodies behave when we cut ourselves. The hormone the plant makes in this case is used in gardening to stimulate cuttings from a plant to produce roots so that they can grow into new plants.

Plants also produce hormones when they are growing under normal conditions. They are responsible for the movement of plants. Single-celled water algae can usually move their whole bodies. Other plants can move only part of their bodies, although fruits, seeds and pieces of plants can be moved from place to place by wind and water currents.

The movement of part of a plant in response to an outside cause is known as a tropism. Geotropism is the response to the pull of gravity. It causes the roots to grow downwards, so they are said to be positively geotropic. Stems grow upwards because they are attracted to the light, and so we say they are positively phototropic. You can see a phototropic reaction if you place a plant in a position where it is partly in the dark. In only a day or two you will notice that the leaves have all turned to face the direction from which the light is coming.

Classification and Names of Plants

Plants are classified according to particular features they have in common with other plants. In the eighteenth century, in the early days of the scientific classification of living things, the principles of classification were details of overall appearance, such as leaf shape and numbers of floral parts. Today the internal structure, the chemical composition and even details of the plant's behaviour are considered. Although the lens and the ruler are still very important tools, the modern botanist might also use a powerful electron microscope and a chemical laboratory.

The basic unit of classification is the species. This is a group of populations of individual plants that have definite and permanent characteristics in common and usually breed with each other to produce similar offspring. Separate species that have features in common with each other are grouped into a genus, and genera (the plural of genus) are grouped into families, families into tribes and tribes into orders. There are also groupings between these categories, such as the subfamily or subgenus. The formation and use of all classifactory names are controlled by the International Code of Botanical Nomenclature.

All botanists are entitled to invent and use their own classifactory scheme and new ones are continually being published. The great majority of these new schemes are really no more than refinements and minor regroupings of older schemes. They all keep the major categories of orders, families, genera, species and so on. These classifications are usually based on theories about the evolution of plants that botanists accept, but cannot prove. They illustrate ways in which plants could have evolved.

However, not all botanical classifications are based on evolution. Plants can be grouped according to the particular and specialized requirements of the classifier. For example, an artist might classify plants according to flower colour, and a gardener according to height, duration (annual, perennial, biennial) and form (herbaceous, tree, shrub, climber). Plant geographers and ecologists frequently use a system based on the position of the overwintering buds in relation to the soil surface.

Attempts have been made to give plants scientific names other than those usually used – even computer digits and numbers in some schemes. Nonetheless, all classifactory schemes, whatever their purpose, refer to species by a double name, or binomial. This binomial system is internationally accepted.

The binomial of a plant is always in Latin. It is given only after there has been a careful and detailed investigation into the plant's relationships to other apparently similar plants and its place in the classificatory scheme. Common names are only used by botanists when they are widely accepted and when they do not refer to more than one species. However, most common names are understood only in a single country or region and can refer to many different plants in different places.

To be strictly scientific the name of a plant should be composed of three parts. The first is the generic name, which refers to the genus. The second is the specific epithet, which distinguishes the different species within the genus. Together these two names form the binomial. The third part is the name of the botanical author who first correctly described the species. For an example, let us take *Achillea millefolium* L., the Common Yarrow or Milfoil, which is a European native weed and is cultivated as a garden plant in Europe, North America and Australia. *Achillea* is the generic name. It is given to all plants considered similar to the Common Yarrow, such as *Achillea ptarmica* L. (Sneezewort or Bachelor's Buttons) and *Achillea filipendulina* (the Fernleaf Yarrow). Most generic names are Latin or Greek in origin. They can be based on any real, mythical or imaginary name in any language, but then they must be Latinized.

Millefolium is the specific epithet that distinguishes the Common Yarrow from other species of *Achillea*. The specific epithet is usually based on a characteristic of the plant, such as its flower colour (purple, *purpurea*), size (large, *gigantea*), habitat (water, *aquatica*), or geographical location (America, *americana*). It can also be derived from the name of a person connected with its

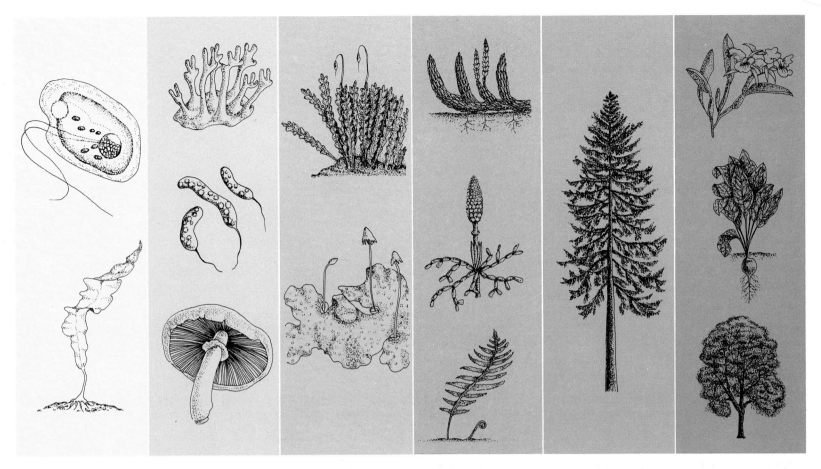

Plants can be classified into groups according to how complex their bodies and reproductive systems are.

This chart shows examples of simple plants at the left, and increasingly more complex ones to the right. The simplest plants are the algae.

Fungi and bacteria are more complex, then come mosses and liverworts, ferns and related plants, conifers and, finally, flowering plants.

discovery or study (Mackay, *mackayi*, or Rothschild, *rothschildiana*). In this case the reference is to the plant's finely cut leaves: *mille*, thousand, and *folium*, leaf.

The third part of the plant's name, that of the botanical author, is abbreviated. L. is the recognized abbreviation of Carl Linné, the eighteenth century Swedish botanist who devised the single, international system of scientifically naming plants and animals that is used today. Other abbreviations are not so short. For example, Wahlenb. is the shortened form for Wahlenberg. Ait.f. is the abbreviation for Aiton *filius* (Latin for son), and refers to William Townsend Aiton, the son of William Aiton. Sometimes two authors or groups of authors are mentioned, the first one in brackets and the second following it: *Dactylorhiza majalis* (Rchb.) Hunt & Summerh. The author in brackets, Reichenbach, first described the species as belonging to one genus, in this case *Orchis*. The authors outside the brackets, Hunt and Summerhayes, first transferred it to its correct genus, *Dactylorhiza*. In general horticultural, botanical and ecological usage the names of the author are usually omitted, but they are always used in scientific works.

Botanists sometimes use additional names for distinct populations, such as varieties and subspecies. Horticulturists, or gardeners, use names that are non-Latin in form and enclosed in single quotes (for example, *Acer davidii* Franch. 'George Forrest') for cultivated forms.

The rules governing the formation of names, the transferral of names if a plant is reclassified and the conservation of old-established and widely-used generic names are periodically revised at special sessions of the International Botanical Congresses every five years.

Bacteria

Bacteria are present everywhere, in the soil, in water, as spores in the air and in the bodies of dead and living plants and animals. We usually think of bacteria as germs that cause diseases, and much of the research into the habits of bacteria is aimed at finding ways of curing and controlling such human diseases as tetanus and tuberculosis. Many bacteria-destroying drugs, which are known as antibiotics, have been found. The most effective of these is penicillin, which is produced by another plant, the fungus called *Penicillum notatum*. Not all bacteria that grow in living animals cause diseases. Several species that live in the intestines of animals are necessary to change the foods the animals eat into forms that their bodies can use.

Most bacteria are microscopic, single-celled plants that multiply by dividing into two identical cells. These new cells grow very quickly until they, in turn, divide into two. This reproduction is so fast that a single bacterium can produce over 32 million similar bacteria in twenty-four hours! However, this growth of a bacterial colony depends on the correct temperature. Most species will grow at temperatures between 20°C and 40°C. Some species that live in the soil in high altitudes will grow at temperatures as low as 5°C. Hot water springs and even the hot water outflow from some factories and electricity

Sausage-shaped bacteria are the most common of all types of disease-producing species. These have been magnified several hundred times so that we can see them.

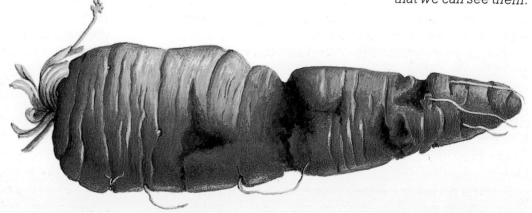

Carrots can be badly damaged and made unfit to eat when they are attacked by bacterial diseases.

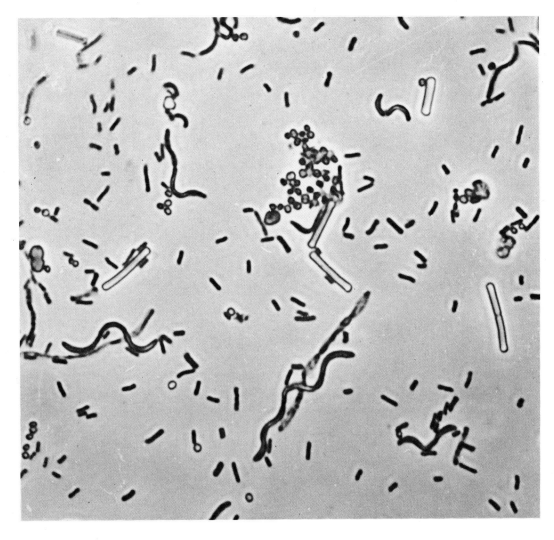

Bacteria form chains of individual plants when conditions are favourable for reproduction.

generating stations provide homes for a few specialized species that grow at 75°C. When conditions are not suitable for growth, when there is little food or water, for example, bacteria produce thick-walled spores that can withstand very severe conditions. In this form bacteria can remain dormant for several years until conditions again become suitable for growth.

Most bacteria are ball-shaped. They can be grouped in pairs, fours, chains, spirals, or irregularly. Some species are shaped like rods, straight or curved. They can be joined end to end to form long threads. Most bacteria cannot move. There are a few types, however, that have one or more little whip-like projections called flagella that they use to push themselves through the liquid in which they live.

Bacteria do not make their own food by photosynthesis. They take it parasitically from other living plants and animals, or from dead organisms. A few types are unique among living organisms because they can exist on inorganic substances like nitrogen, iron, sulphur and ammonia. When the Earth first cooled down, the atmosphere contained a great deal of ammonia and it is possible that bacteria that use this chemical as their food were among the first living things. The nitrogen bacteria are very important because they cause nitrogen and oxygen in the air to unite chemically to form nitrates, which are essential to the life of flowering plants. Some of these nitrogen bacteria live in the soil and others live in the roots of plants, where they cause the roots to swell and form rounded lumps called nodules.

The bacteria that live in the soil are very important to plants and to people. They help to decay dead material by turning it into simpler substances that can be used again by living plants. Some of these bacteria are used to break down solid sewage and turn it into harmless chemicals, and others can be used to purify water.

Viruses and Phages

Viruses are parasites on living plants and animals that cause many serious diseases. Crop plants, domesticated animals and humans all suffer from virus diseases. Smallpox, influenza and poliomyelitis are all human illnesses caused by viruses. Viruses are so minute – about a fifteen-thousand millionth of a millimetre long! – that they can be seen only with an electron microscope. They can pass through the finest filters scientists have developed. It is almost certain that they would not have been discovered except for the fact that they produce diseases.

Many scientists have argued that viruses are not really plants or animals, and that they are not living because they can be crystallized and redissolved in the same way as ordinary chemicals. However, viruses can reproduce themselves and pass on their inherited characteristics, which only living things can do. The viruses are often considered as plants because they are somewhat similar to small bacteria.

The individual virus is very simple in structure – it is not much more than a few complex protein molecules. When these invade a cell of the living host plant or animal, they make it produce more viruses instead of more host cell material. The resulting virus offspring then break out of the host cell, thereby killing it. The rate of reproduction is very fast, and a whole plant or animal can be killed by a virus in less than one day. The protein molecules in a virus are very similar to those found in all the cells that carry genetic information from one generation to another in living organisms. It is believed that many forms of cancer may be caused by viruses. It is possible that the virus enters a host cell, where it becomes part of the cell's own genetic material. Then it may become active after lying dormant for many years.

The virus diseases of some plants can get passed on from one generation of the host plant to another. The colour markings of the flowers of Rembrandt tulips are

Even greatly magnified, virus particles would look only like these rods. A single virus would have to be enlarged still further for us to see the sausage-shaped individual molecules.

Colonies of phages resemble tiny living organisms, but their regular crystal-shaped bodies suggest that they are chemical substances.

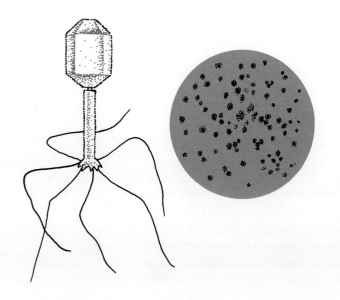

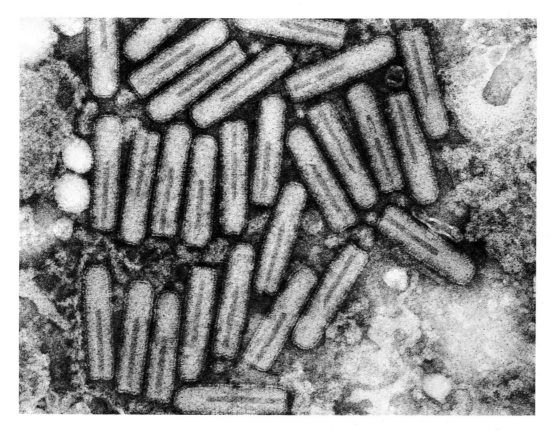

Viruses cause diseases in many plants and animals. These virus particles have been taken from the sap of a diseased brussels sprout plant and have been magnified almost 100 000 times.

caused by an inherited virus disease.

Although bacteria are very small, viruses are very much smaller and can even infect bacteria. These specialized viruses are called bacteriophages, which is usually shortened to phages. The phage attaches itself to the cell wall of a bacterial cell and injects its contents. As with other viruses, the bacteriophage causes the bacteria to produce more phages and not more bacteria. Eventually, the bacterial cell bursts and releases the phages, which immediately attack other bacteria nearby. Bacterial diseases can be controlled by introducing a phage produced in a laboratory into the diseased organism.

Bacteriophages enter the body of a bacterium and cause it to produce more bacteriophages rather than more bacteria. Eventually the bacterium dies and the new bacteriophages are released to invade more bacteria.

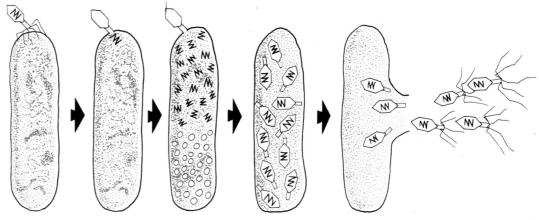

The Algae: Seaweeds

Algae, together with bacteria and fungi, represent the earliest forms of life on Earth. Fossilized algae dating back more than three thousand million years have been found in some rocks and are the oldest known fossils. Most algae are aquatic and are found in the sea either as plankton or as seaweeds.

There are various ways of classifying seaweeds. One useful classification, which is also based on technical characteristics, is into colour groups. So at a glance we can tell whether we are looking at the red algae, the brown algae, or the green algae. The green species grow where there is most sunlight just under the surface of the sea. At lower depths, where only the green and blue light rays of the Sun penetrate, the brown and red algae thrive because they use these light rays to make their food.

Seaweeds are found throughout the world's oceans and, because they are most plentiful near the coasts, they have been used by people in many ways for thousands of years. Many seaweeds are rich in iodine and can be used as food for grazing animals such as sheep. Some kinds are

Seaweeds are found in large numbers at low tide level. Brown seaweeds can withstand being exposed to the air when the tide recedes, but red seaweeds quickly become dehydrated and die.

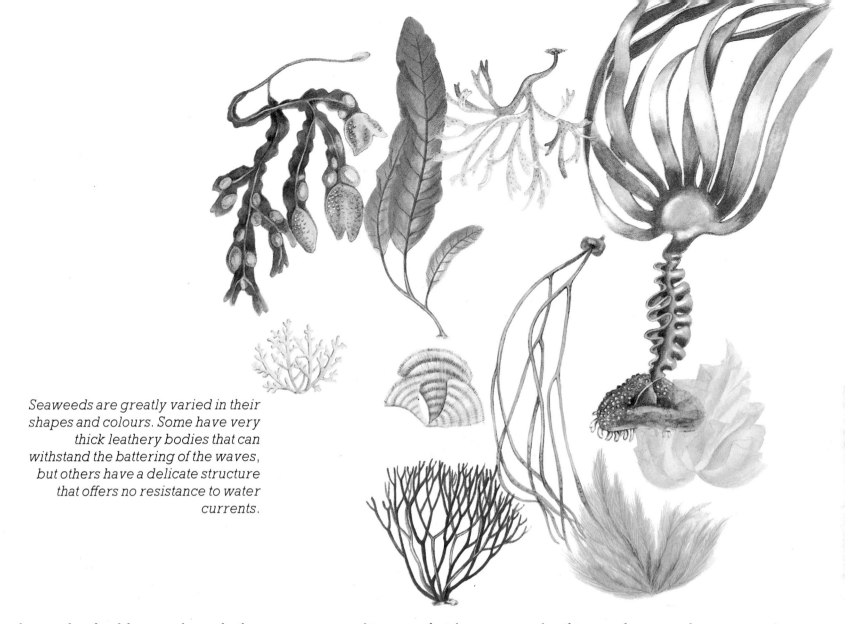

Seaweeds are greatly varied in their shapes and colours. Some have very thick leathery bodies that can withstand the battering of the waves, but others have a delicate structure that offers no resistance to water currents.

also used as food for people, and others are processed to produce chemicals that give bulk and a smooth texture to factory-made foods such as ice cream. When burned, seaweeds yield a fine ash that is rich in potassium salts, which is a good chemical fertilizer for the land. The best seaweeds for making fertilizer are the kelps, some of which are very large. One species, the Giant Kelp, grows deep down in the Atlantic Ocean, sometimes as much as 30 metres below the surface of the water. This species can produce a repeatedly branched stem up to 174 metres long, making it one of the largest plants in the world.

Most seaweeds reproduce sexually. It is a fairly simple process, although it varies in detail from one species to another. Special cells in the body of the seaweed produce male and female cells. These cells can be very simple and superficially identical with other, ordinary body cells, but in some species they are larger and more complex. As in other types of plant, the female cell is fertilized by the male and the result is the formation of an embryo, which soon develops into another plant.

Seaweeds are simple, usually feathery or ribbon-like plants that have no system of veins and no roots, leaves or woody parts. Instead of a root, many of them have a special organ called a holdfast. By means of its sucker-like outgrowths, it enables the seaweed to remain anchored to rocks during the roughest seas. If for any reason the holdfast breaks away from the rock, the seaweed will drift away and die. Some species do not have a holdfast and live by floating freely through the water. One of these free-floaters is *Sargassum natans*. It does not seem to reproduce sexually. New plants grow from fragments that are detached from the body of the plant.

The Algae: Diatoms and Desmids

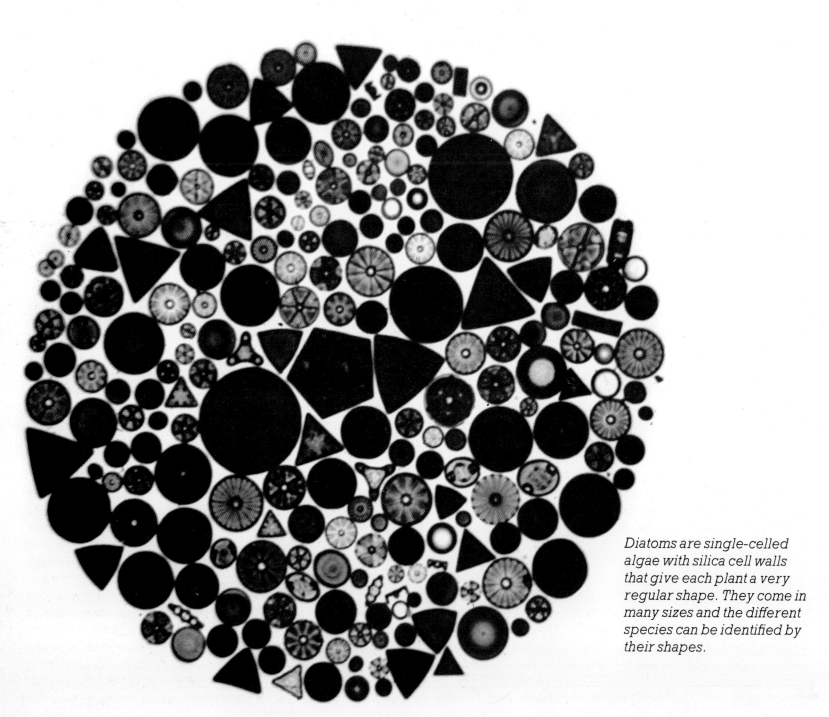

Diatoms are single-celled algae with silica cell walls that give each plant a very regular shape. They come in many sizes and the different species can be identified by their shapes.

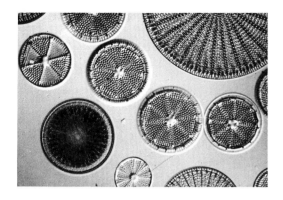
The delicate markings on the silica cell walls remain even when the plants are dead and turned into fossils.

Although they look almost mechanically shaped, diatoms are living plants that grow in water.

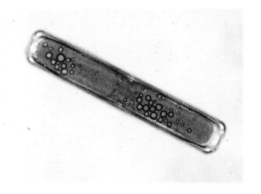
Living diatoms, such as this freshwater species, can accumulate oil in their cells, which remains when the rest of the cell has died.

You will never notice a diatom when you are swimming in the sea, but if you were to look through a microscope you would be amazed by the beauty and variety of these minute plants that live in the water. The most characteristic feature of the diatoms is that the outer layer or wall of each cell is impregnated with a hard mineral called silica. This gives each species a very distinct appearance when viewed through a microscope. The stiffened cell wall is like a box in two almost equal parts that overlap. This is why these algae are called diatoms, meaning two atoms. The silica on each cell wall is arranged either in a circular, radiating pattern, or in two parts side by side. The silica cell walls remain when the living algae plant inside dies and decays, and they can become preserved in the deposits that form at the bottom of the pond or sea in which they live.

Nearly 6000 species of diatom are known. The many living species are an important source of food to fish and animals. Some of the diatoms are known only as fossils however. Where conditions for diatom growth were extremely favourable in the past, such as when the seas were still very warm, vast deposits of diatom fossils were also formed. Many of these 'diatomaceous earth' rocks lie close to deposits of oil. It could be that oil was formed from plants that were very similar to diatoms but did not have the rigid silica cell walls. In the Santa Maria oil fields in California deposits of fossilized diatoms over 1000 metres deep have been discovered, and more than a million tonnes are processed for use every year.

As a result of the exploration for oil much has been discovered about diatomaceous earth and the ways in which it can be used. Scientists soon found that this very fine material was also very absorbent. They used it to absorb a highly dangerous liquid explosive, nitroglycerine, and in this way made the easily transportable solid explosive called dynamite. As well as being used as a filter in many industrial processes, diatomaceous earth is being used more and more as an insulating material because it does not shrink, melt or disintegrate at temperatures even above 500°C and is not harmful to people.

The silica of the cell wall is a very hard material and so these fossilized plants are also sometimes used as a very fine abrasive and polish for metals and softer rocks.

Desmids are very similar in appearance to the diatoms. However, whereas the surface patterns on diatoms are made from silica, in desmids they are spiked and made of very hard cellulose. In some species the spikes are further strengthened by iron compounds.

The cells in desmids can be solitary, joined end to end to form hair-like threads, or stuck together in shapeless colonies. In nearly every species the cells are squeezed together in the middle, dividing them into two distinct halves, which are joined by a connecting zone.

Many desmids are able to move, not by the little whip-like projections used by bacteria, but by means of a series of jerks. These are probably caused by the expulsion of a jelly-like substance from one end of the cell.

The Algae: Freshwater and Terrestrial

Most algae are much smaller than the seaweeds and, indeed, many of them are among the world's smallest green plants. Many algae are single-celled and often exist as individual plants, but some of them live in groups. Other algae are threads of cells and form thick hair-like mats in ponds and lakes.

The majority of the single-celled species live in the sea, usually in the upper layers. There they form an important part of plankton, the drifting mass of minute plant and animal life that is so important as a food for many species of fish. The algae in the plankton are called phytoplankton, and are the starting point of all marine food chains.

Millions of microscopically small single-celled green algae plants can be found on damp tree trunks even in heavily polluted inner city parks.

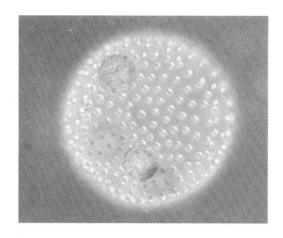
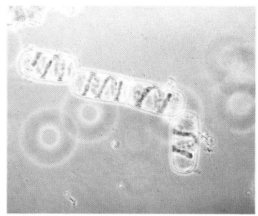

Both these species of algae live in water. Those on the left are single-celled Volvox, *and those on the right are the thread-like* Spirogyra *that are found in every ditch and pond.*

Many other single-celled algae live in freshwater, such as those that form the familiar scum on ponds. A few single-celled species live on land. For example, there is one species that forms a greyish-green powder when dry, and a bright green one when damp, on tree trunks, wooden fences, brick and stone of all types.

These freshwater, marine and terrestrial algae are classified into five main divisions. Members of the many families of the *Chlorophyta* division range from single-celled land forms to large seaweeds. All species make carbohydrates from carbon dioxide and water by photosynthesis, and are among the most important producers of food for animals. They are colonizers of suitable habitats and, as they can reproduce both sexually and asexually very rapidly, they can exist in enormous quantities, with tens of thousands of individuals of one species closely packed together.

The *Euglenophyta* algae are very unusual in that most of them also have been claimed by zoologists as animals! This is for three reasons. Firstly, they can move because they have one or more whip-like flagellae, which they use to propel themselves through the water in which they live. Of course, bacteria can do this too, so it is not a good enough reason by itself. Secondly, they have a light-sensitive red eye spot, which enables them to seek out the light so that they can photosynthesize at the fastest possible rate. Thirdly, many of them can eat solid foods. However, it is because the *Euglenophyta* produce food by photosynthesis like other green plants that most scientists consider them to be plants. Several species produce very thick-walled cysts, which are a prolonged resting phase. This mechanism for remaining inactive also indicates that they are plants rather than animals.

The members of the *Pyrrophyta* division are similar to those of the *Euglenophyta* in that most species have flagellae and possibly take in solid food. They differ from them in that their colouring pigments are not typically grass-green, but vary from yellowish-green to golden-brown. Most species are single-celled and some, such as *Dinamoebidium varians*, are very similar to animals like the amoeba. Another group have their cells joined end to end to form branching threads.

The *Cyanophyta* algae usually have a blue colouring pigment that can mask the green chlorophyll. This is why they are called the blue-green algae. This division is unique in that sexual reproduction has never been observed. All multiplication takes place by cell division. The resulting offspring usually stay together to form threads or a colony held together in a jelly-like envelope. Some species can move by sliding along on a slime that they produce.

Most species live in freshwater, but some grow on damp rocks, in and on soil as colonizers and a few grow in the outflow from hot springs. These thermal algae, as they are called, can even multiply in temperatures as high as 75°C. It has been suggested that they were among the first plants in the world and lived in the hot waters that were much more widespread when the Earth first started to cool down.

The 6000 species of the *Chrysophyta* division include the yellow-green algae, the golden-brown algae and the diatoms. The yellow-green and golden-brown species can be solitary or united into colonies. They live in the sea, in freshwater and in damp spots, where they are often intermingled with mosses, liverworts and other kinds of algae.

The Fungi: Moulds and Yeasts

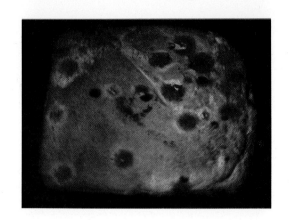

At some time you have probably seen a piece of bread that is very stale covered with small bluish growths. These moulds are one of the many species of fungi. Fungi (the plural of fungus) are simple plants that, like the algae, are not divided into roots, stems and leaves, but have a body. This body consists of very fine branching threads, called hyphae, that are actively feeding and growing. Rather like a cobweb, they form a network over, and usually also in, the food on which they live. Sometimes the hyphae are divided into cells, but often there are no dividing cell walls.

Fungi do not have chlorophyll in their cells and therefore do not make food by photosynthesis. Some, called saprophytic fungi, take their food from dead and decaying plants and animals. Parasitic fungi get their food from the tissues of living plants and animals. These fungi help to maintain the balance of nature in two ways. Parasitic fungi cause diseases and in this way control the numbers of most living things. Saprophytic fungi help to break down dead bodies, leaves, old wood and waste matter from plants and animals into simpler chemicals that can be re-used by plants.

However, saprophytic fungi can also be harmful and destructive. Several species grow on leather and natural fibres such as cotton. Dry rot fungus can attack wood, and is a real nuisance in old buildings. Some fungi attack prepared foods of all kinds, others can live on petrol and reduce its ability to drive engines. Still others attack paint,

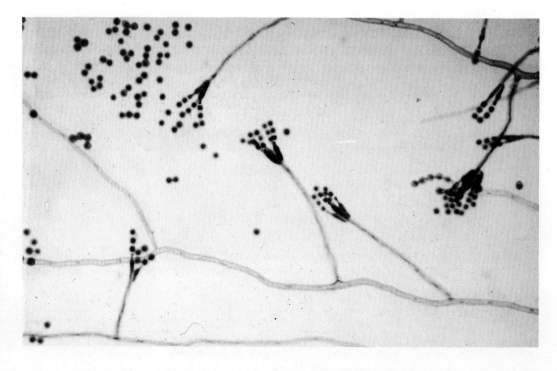

The spores of a Penicillium *fungus are released from the tips of special long spiral threads. These spores can remain alive for many years, existing in the soil and in the air. They will germinate only when the right food is available.*

Fungus will grow on many different kinds of food. Damp bread and damp, cooked rice, if left in a room for a week, will become covered with colonies of fungus. The brown rot seen on fruit, such as this apple, is also caused by a fungus.

and many can quickly destroy stonework and bricks.

There also are useful fungi. For example, some moulds are used as flavourings in blue-veined cheeses, and the mould *Penicillum notatum* is processed to make the antibiotic wonder-drug penicillin.

Fungi can reproduce sexually by the fertilization of sex cells, or asexually by the production of spores. Fungus spores are very small – in fact, one thimbleful of soil can hold 100 000 living spores.

Fungi are divided into six classes. The slime fungi live on decaying plants and animals. Their slime is a jelly-like mass of living protoplasm that moves slowly as it feeds. Eventually, it takes a more definite shape and produces spores.

Bacteria are really a very specialized form of fungi, and the *Actinomycetes* are very similar to them. In both these classes of fungi the hyphae do not form networks. Many of these species produce useful antibiotics.

Another class of fungi includes food moulds and parasites of insects. Some species in this class look very much like the algae, and many of them live in water and are parasites on fish.

The fifth class includes the rusts and smuts that are parasites of plants, and the mushrooms and toadstools.

The last group also includes some plant parasites, as well as moulds and yeasts. Yeasts are single-celled fungi. They can convert, or ferment, sugar into alcohol and carbon dioxide, and are used in brewing and baking.

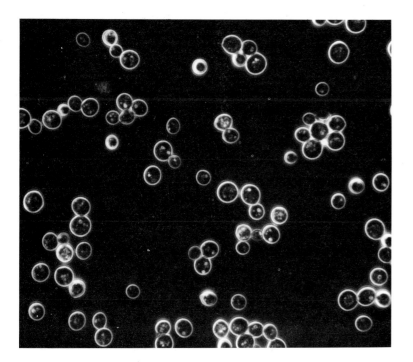

The microscopically small single-celled yeast fungus rapidly forms large colonies of grouped cells when there is the right amount of food and warmth.

The Fungi: Mushrooms and Toadstools

To many people mushrooms are large fleshy fungi that can be eaten, and toadstools are large poisonous fungi. However, mushrooms and toadstools include all the large fungi with a conspicuous spore-producing body.

A true mushroom, such as the edible, cultivated field mushroom we can buy in shops, is the fruiting body of a soil-dwelling fungus that pushes its body up through the soil or special compost when atmospheric conditions are damp and not too cold. The cap and stalk of the mushroom are a mass of closely-packed hyphae. Hyphae on the underpart of the cap form the delicate, thin gills. The spores are produced in large numbers on the surface of the gills.

Toadstools, both the harmless and the highly poisonous species, are very similar in structure to the edible mushrooms. They are distinguished from mushrooms by differences in such features as the shape of the gills, the colour of the spores, and the colour and markings on the skin on the top of the cap. In many of the poisonous species of toadstools this skin is brightly coloured, such as the scarlet-capped Fly Agaric. However, colour must never be taken as the only guide as to whether a fungus is safe to eat. The most poisonous European toadstool, with the striking name of the Destroying Angel, is pure white and very, very similar, especially in its young stages, to the field mushroom.

Some fungi are edible (left), others are poisonous (centre), and some destroy the wood on which they grow (right).

Fungi can look like little birds' nests, deer's antlers, or dried-up pieces of wood. These fungi feed on dead and decaying wood and humus.

Toadstool poisons are complex chemicals that affect the chemistry of the animal that eats them, and often lead to death. However, some animals can safely eat certain toadstools that would be deadly to other animals. The nature and action of many toadstool poisons is not yet understood by scientists.

A stroll in a damp woodland or grassy field in the autumn will reveal many kinds of large fungi. Not only ordinary stalk and cap fungi will be seen in great quantity, but many others that look quite different. There is one species that looks like pieces of discarded orange peel, and another that resembles skinned rabbits' ears. You might just mistake some fungi for cauliflowers.

Others look like miniature versions of reindeer horns, and still others like jugglers' clubs. Puffballs look like large brown or grey pebbles. When they are kicked or stepped on, they emit a puff of khaki spores. There is one species of puffball that can grow to more than 75 centimetres wide, and more than once a colony of them has been reported to the police as objects from outer space.

Probably the most unusual fungi are the Bird's Nests. They appear as a thimble-sized brown cup on rotting wood. At first the cup is closed, but when it opens the spore containers inside look exactly like little white bird's eggs. Related to these are the Earth Stars. Sections of their ripe round fungus body fold back and press against the ground like a star. In country districts in earlier times legends grew up around these fungi, which were believed to have arisen from dust shed by shooting stars.

Even if they are not seen, the Stinkhorns are easily detected by their overpoweringly offensive smell. The hyphae growing underground produce a 'witch's egg', which soon bursts and releases a long white horn with a slimy olive-green tip carrying the spores. The rotting-flesh smell soon attracts blowflies, which feed on the slime and carry away the spores.

The Parasol Toadstool is very common in grassy areas and the edges of woodlands. The top opens out just like a parasol as the toadstool grows.

Lichens

Certain plants are able to live together in a special relationship in which each species provides something such as food, water or protection that the other cannot provide by itself. This type of relationship that benefits both plants is called symbiosis. A good example is the combination of a fungus living in the roots of a forest tree. It passes water and minerals from the soil to the roots, and takes food from the tree in return.

Lichens are composite plants in which a certain type of fungus and a single-celled green or blue-green algae live together, both benefiting from the association. The fungus obtains food from the algae, and the algae obtain mineral nutrients, water and protection from over-exposure to sunlight and winds from the fungus. In this way lichen can exist where neither the fungus nor the algae could live independently.

The body of a lichen can be like a thread, a jelly, a crust or a leaf. It can be upright or dangling, branched or unbranched. Some species are very small, like the black encrusting lichen found on tide-level shingle, which is

The lichen Usnea comosa *grows all around the older branches of trees in the damper parts of England, but some species, such as* Parmelia physodes, *live in isolated patches on the bark of the main trunks.*

The Pixie lichen grows on very acid soils on heathlands everywhere. These lichen were photographed on a heath in Alaska.

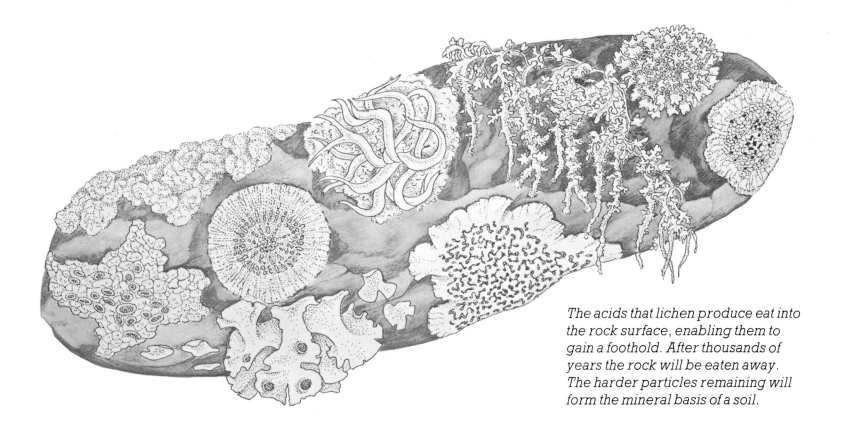

The acids that lichen produce eat into the rock surface, enabling them to gain a foothold. After thousands of years the rock will be eaten away. The harder particles remaining will form the mineral basis of a soil.

smaller than the head of a pin. Others can form twisted masses of branched threads over one metre in diameter. In some lichens the body is simply the hyphae of the fungus in which cells of algae are entangled. The more complex lichens have alternating layers of fungi and algae, like a sandwich. Some of the hyphae in the lowest layer of fungi act as roots, anchoring the lichen to the soil, rock, leaf or tree-bark on which it lives.

Lichens reproduce in one or more of several ways. Small broken-off pieces of the body can grow into new plants, or masses of algae cells and fungi hyphae can group to form special bodies that are shed from time to time. They can reproduce sexually too by means of fungus type spores, which germinate and form a lichen if the right alga is present.

Lichens are found in every country in the world, and although only 16 000 different species have been recorded, millions of individual plants of any single species can grow in a small area. All kinds of situations are suitable for lichens, from the leaves of tropical rain forest trees to the otherwise bare, frost-shattered rocks of the polar regions. In the Antarctic lichens are by far the most diverse and abundant plants, and in the Arctic tundra the reindeer moss, the principal food of reindeer, is really a lichen. In Iceland people eat a local species, and in many parts of the world lichens are collected and boiled with fabrics, especially wool, to colour them.

Lichens contain many substances to produce these dyes and some lichens can be identified by noting the colour they turn when certain chemicals are dropped on the body. Litmus paper, which turns red in acids and blue in alkalis, is made from lichens and is used throughout the world in laboratories and in industry.

Lichens are very strong and can survive extremes of climate. They colonize bare rocks, concrete, bricks, asbestos roof-sheeting and even glass, breaking all these materials down into small particles to form a soil. One thing that they cannot tolerate is air pollution. Around nearly all major industrial and residential cities and large towns chemical-laden smoke from factories and houses has killed most lichens. Some species are more resistant than others and by counting the number of species within certain distances of the city centres it is possible to build up a map of pollution concentration.

Ferns and their Relatives

Fern leafs, or fronds as they are called, are often divided into many little leaflets, each with its own veins and spore capsules.

The frond of the Male fern just beginning to unfurl.

Ferns and their relatives are simple plants that do not possess flowers and reproduce by spores rather than seeds. Although there are several other groups of flowerless plants, such as mosses, fungi and algae, ferns are unlike them in that they have a system of veins and strengthening tissues somewhat similar to those in flowering plants.

Ferns are found wherever there are moist conditions. Typically, this includes forests and damp, marshy areas, but several species will grow rooted in crevices on bare rocks if the air is humid. There is also a group of ferns that lives on or in water, and one of these has become a major pest in reservoirs and artificial lakes in tropical countries. Although ferns require the same conditions for growth as flowering plants, most of them can stand low light levels and make good houseplants.

During the nineteenth century ferns became extremely popular as plants for the home, as well as for growing in greenhouses, conservatories and outdoors. Special closed glass cases, like miniature greenhouses, were made to give the ferns the damp atmosphere they need. This Victorian fern craze, as it was called, involved not only normal ferns but also monstrosities that occasionally occur in the wild. Forked and crested and other distorted leaves became more desirable to the grower than the ordinary and more graceful forms.

In the southern hemisphere some ferns have persistent leaf bases that remain when the fern leaf has died and withered away. These leaf bases eventually form a trunk-like stem and the whole plant, called a tree fern, looks like a small palm tree. In prehistoric times, before there were any flowering plants in the world, tree ferns were very common. Today they are quite rare, and can be found only in New Zealand and South America.

Closely related to ferns are the clubmosses and horsetails that were the dominant plants millions of years ago. Very few species of these primitive plants are found today, but a few have managed to compete with flowering plants and have become weeds, such as the Field Horsetail. Clubmosses occur in both temperate and tropical regions, and are most spectacular in the tropical rain forests,

There are many different types of fern. At the top, the fronds of the Hart's Tongue are almost completely unfurled. Above is bracken, which is found in woods, grasslands and moors in many countries. It spreads by means of long, underground rhizomes.

where they festoon the lower branches of the trees.

There are several other groups of fern allies. One family alone has about 700 different species, most of which grow in the tropics, although there also are temperate species. *Selaginella kraussiana* belongs to this family. A trailing plant, it is a native of Africa that now grows in the warmer parts of Europe and is very commonly found in greenhouses.

The Quillworts, which are completely aquatic, occur in all parts of the world, but are rarely seen, and other plants related to ferns grow on damp rocks or the bark of trees in the tropics and subtropics.

Mosses and Liverworts

Mosses and liverworts are small non-woody and flowerless plants that do not have a well-defined system of veins. They can grow in all kinds of habitats from the dry tops of brick walls to the forest floor, and from the branches of high altitude rain forest trees in the tropics to bogs and fens in the Arctic and Antarctic tundra. Many species are very tolerant of drought. They can seem to be dead after a long hot, dry season, but as soon as they are moistened, they become green again and continue their growth. This ability to survive in harsh conditions means that they can be among the first colonizing plants on newly exposed rocks and other fresh surfaces. In most cities and towns in Europe and North America the small spaces between paving stones are filled with the minute silvery-grey leaves of these daring pioneers.

Although mosses and liverworts are ecologically important because they act as colonizers and prepare the site for larger plants, very few are of use to people. The exceptions are mosses of the genus *Sphagnum*, usually called bog mosses, although some grow in woodlands. Sphagnum mosses are extremely absorbent, and those that live in the wettest areas of bogs can hold many times their weight in water. As the water in bogs is very acid, the water the mosses hold is sterile. During the First World War great quantities of Sphagnum moss were collected and dried and used as sterile, surgical dressings. Today they are used for packing plants for posting and as a water-retaining ingredient in potting composts for many greenhouse plants. The dead and partly decayed remains of this same species are the main ingredient of peat, which is used as a fuel and in fertilizers.

Mosses and liverworts absorb water and mineral nutrients directly through their leaves, as their little hair-like roots are usually only to anchor them to the surface on which they are growing. All mosses have stems, which are up to one metre long in some tropical species, and all have leaves, although they may be very small in some of the dwarf species. Superficially, liverworts look like

Mosses can grow on humus-rich soils, such as is found under trees, where they feed on the decaying leaves. The White Fork Moss forms large cushions, which easily become detached and enable the plant to colonize large areas.

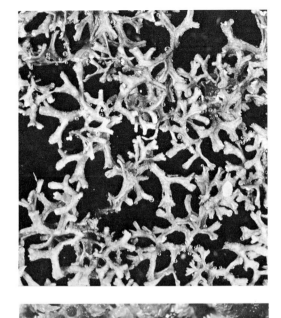

Several liverworts are completely aquatic, such as the Floating Crystalwort, left. Pellia epiphylla, *right, is a species common on damp earth in woods, gardens and the edges of fields.*

The Sphagnum mosses, left, form the main ground cover in acid bogs. Pohlia nutans, *right, grows on dead wood in forests, and can easily be recognized by its long stalked capsules.*

mosses, having stems and leaves. However, the most noticeable liverworts grow as flattish structures with rounded lobes, and look like the flat bodies of brown seaweeds, except that they usually are a bright, shiny green. This type of liverwort has no leaves or stems, and the lobes are held to the ground by the hair-like roots.

Liverworts and mosses are similar to ferns in that they reproduce by the method of alteration of generations. In the case of ferns, however, the two kinds of plants are quite separate, and in mosses and liverworts they grow together. The moss or liverwort plant that we often see growing is the sexual form that produces the male and female cells. When the male has fertilized the female, the fertilized egg develops into a non-sexual spore-bearing plant, that grows on a short stem above the main plant.

In liverworts the spore capsule eventually appears at the top of this plant. When the spores are mature, the capsule bursts and sheds the spores. If the spores fall on moist soil, they will germinate and develop into typical liverwort plants.

Some liverworts can also reproduce asexually. They grow small detachable buds, which are washed out by rain splashes and soon grow into new plants.

When moss spores have ripened, scattered and germinated, they develop into a very fine hair-like green plant, from which clumps of the typical moss plant grow.

Water Plants

All plants require water for their food manufacture and growth. Some, like ferns, mosses and liverworts, live on land and require water for reproduction. Others, like most algae, spend their entire lives in water. Although most flowering plants live on dry or marshy land, there are also thousands of species that remain more or less totally submerged for most of their life-cycle. These are called aquatics, or water plants. The great majority of aquatic flowering plants live in freshwater but some, such as the Eel Wrack, occur in seawater.

Water plants are often very specialized. Most have flexible stems that bend easily in the water currents, and in many species the leaves are finely divided and almost hair-like to reduce friction with the water. The flowers of water plants can be pollinated in the same way as land plants, with the pollen being transferred from flower to flower by insects. However, in some species the pollen is carried by water currents and caught by the trailing underwater stigmas.

Some water plants also produce special buds that break off from the main plant and fall to the bottom of the water, where they remain throughout the winter. These

Many unusual plants can be grown indoors in a tropical aquarium, and are essential to replace the oxygen used up by the fish.

The ribbon-like leaves and female flowers of Vallisneria gigantea.

These four common pondweeds show how varied water plants can be.

buds start to grow in the spring when the increasing daylight turns their stored food supply into oils, which helps them rise to the surface and produce a new plant.

Among the first plants colonizing ponds and lakes are those called 'free floaters'. They are not rooted in the mud at the bottom, but what roots they have dangle in the water. They are at the mercy of surface currents and are a readily available source of food for fish and water birds. Nevertheless, they can colonize enormous areas of water, and in this way can become major pests that prevent the passage of boats, clog up hydro-electric generating plants and, by keeping sunlight out, kill off the animal and plant life in the water and cause pollution. One example of these water weeds is the beautiful blue-flowered water hyacinth, found in America, Africa, eastern Asia and Australia.

Other well-known free-floating plants are the duckweeds, which probably are the smallest flowering plants in the world. They consist of a very small floating or slightly submerged green pad, which behaves like a leaf. Dangling down from this is a long root or a bunch of rootlets. The flowers, which are generally too small to be seen by the unaided eye, grow from the pad.

Most water plants are rooted in the mud or attached to the boulders or gravel at the bottom of a pond, lake, stream or river. Their leaves can be totally below the water level at all times, but many plants produce solid floating leaves as well as, or instead of, the submerged feathery ones. Some water plants that have their roots and main stems below the water, have their leaves above the water surface. This type also may or may not have submerged and floating leaves as well.

The type of leaf or leaves produced by a water plant may be fixed, but several species can produce different leaf types on the same plant as the water level changes. When the water level falls in times of drought, some plants can survive on the drying mud, producing more or less ordinary land-plant type leaves. Other water plants, although capable of producing floating or above-water leaves, often will not produce them when the water speed is too great. Then they exist only on their underwater leaves.

Marshes, Bogs and Fens

When the water level is at, or just above, or just below, the soil level the only types of plants that will be able to grow are specialized ones that do not mind a waterlogged soil. If the soil and water are acid and poor in minerals, the vegetation that will develop will be called bog or fen. Fens and bogs are similar, but fens have more minerals in their water, and so can support more plant species.

Bogs and fens only occur where the rocks under the soil allow acid water to accumulate. This prevents the total decay of dead plants. The partially decomposed remains of bog and fen plants are preserved by the acid water, just as onions and other vegetables are pickled by vinegar. Each year a new layer of preserved plants is added and the lower layers become very compressed. The total resulting mass of plant remains is peat, the upper layer of which acts as the soil for living plants. Peat is very poor in plant foods and therefore bog plants are either mosses, which require very small amounts of minerals, or flowering plants that get the minerals they need from sources other than the soil or the water. Many bog and fen plants, such as heathers, alders and birches, have swellings on their roots. These are caused by special kinds of bacteria that help the plant by making essential nitrates from oxygen and nitrogen in the air.

Other bog plants attract and catch insects, digest their bodies and then absorb the nitrates and other chemicals produced. In some plants the insects are caught in pitcher shapes that grow at the ends of the leaves. The pitchers are filled with juices in which the insect drowns and which then help the plant to digest its body. Other plants have sticky glands on the leaves that entangle the insect, and the Venus Flytrap has a pair of leaves that spring together when an insect touches them.

Marshes usually occur on soils rich in minerals and with plenty of decaying plant material. The water in marshy soils is not acid, and so peat does not form. Swamps are a form of marsh that develop in places where the water level is up to one metre higher than the soil for at least one-third of the day.

Marshes have many more species of plants than bogs, fens or swamps. Marsh plants do not need special mechanisms to enable them to obtain nitrates, but the waterlogged soil means that the roots of plants growing in it can be deprived of air. Many marsh dwellers have developed

Marshes are formed by the silting up of long stretches of open water. Here a ditch has become filled with pondweed, Frog-bit and Arrowhead.

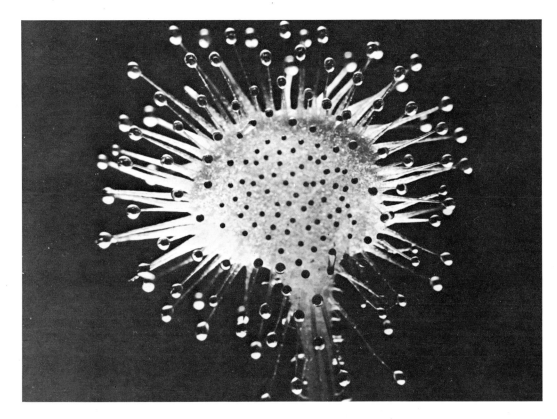

The Sundew, above, captures insects with its sticky-tipped hairs. The leaves of the Venus Fly Trap, below, snap shut when an insect touches them.

Marsh orchids are very noticeable for their large, colourful groups of flowers, or inflorescences.

large internal air spaces that connect the stems and roots and in this way provide air from above ground so that the roots can function.

A marsh can develop wherever mineral-rich water can accumulate. This may be around lakes and ponds where sand and silt gather and raise the underwater soil level so that it eventually reaches water level. It can also happen where spring water seeps out of rocks on a hillside and keeps the soil wet. In the first case the marsh may only be temporary because the conditions are suitable for tree seedlings. After many years the seedlings may grow into large trees that overshadow the original marsh plants and dry out the soil by transpiration. The great volume of leaves produced by trees adds to the soil. Eventually, the soil level will be permanently above the water level and a woodland will exist in an area that was originally open water. Marshes forming around springs are permanent because the slope does not allow soils to develop that are deep enough for mature trees.

Mountain and Seaside Plants

Although the tops of the highest mountains are covered with snow and ice all the year round or are so cold and windswept that no plants can grow, most other upland areas have their own plants. It is not possible to state at what height upland plants are to be found, as this varies greatly depending on factors such as the nearness of the sea, the distance from the equator, the kind of rock and soil and the wind pattern of the area. The amount of direct sunlight can be important too. In the northern hemisphere typical upland plants first occur higher up the sunlit south side of a mountain than the shaded and cooler north side. Upland vegetation occurs wherever the temperature and conditions for lowland plants are not available.

Mountain plants have to withstand harsher conditions than other plants. They survive because they have become adapted to lower temperatures, high wind speeds and often poorer soils. Many mountain species are covered with ice and snow for much of the year. They come into leaf, flower, produce their fruit and shed their seeds only in the short period after the winter snows have melted and before the first autumn frosts occur. Some plants are able to survive very low temperatures even when in flower. Arctic Scurvy-grass plants have suddenly been overtaken by winter conditions when they were in full flower and continued to complete their life-cycle six months later when the temperatures rose again!

Mountain plants are often called Arctic-alpines because they also occur either in the Arctic circle or in the Alps in France and Switzerland. Rock garden plants are often mountain species. They do not always survive in gardens at low altitudes because the winter conditions are not cold enough for them to rest, and the soft growth they produce is not resistant to damp air and heavy rainfalls.

Near the sea the air temperatures are not as extreme as they are further inland, being cooler in the summer and warmer in the winter. This means that many plants can grow near the sea that could not survive elsewhere. Seaside plants have to withstand salts from the sea, and many inland plants cannot grow near the sea because of this salt. Seaside plants therefore can be quite distinct from others in a country.

In the hotter parts of the world trees grow very close to the sea and the trees of mangrove swamps grow with their roots in the mud and sand below high tide level.

Although many mountain plants are often picked and dug up, the absence of heavy grazing, mowing and ploughing has enabled several species to survive. The grasslands and rocky ledges above the tree limit in the Alps are often full of flowers.

Sea cabbages are some of the plants scattered among the large pebbles on a shingle beach.

Marram grass can continue to grow even when it is covered by sand. It is used to stabilize sand dunes because its network of roots holds the sand together.

However, in the temperate regions, such as most of Europe and North America, the combination of salty air and high winds at sea level means that most trees, if they can grow at all, are very stunted and misshapen.

The soils found near the sea, as well as being full of salt, are usually made of rock particles that are all of one type, having been sorted out by the action of the wind and the sea. The most unusual seaside soil is shingle – large stones up to 35 centimetres across that accumulate in long ridges just above high tide level. These shingle beaches have no plants on the side facing the sea because the force of the waves hitting them would wash any plant away. On the side facing the land, however, sheltered from the highest winds and waves, lichens, mosses and many flowering plants are found. Although the heaped stones are so large and have large air-spaces between them, their surfaces facing inside the pile are always wet with *freshwater*, not salt water! This is moisture from the damp sea winds that has condensed on the stones' cold surfaces. The roots of the plants use this water and obtain mineral nutrients from decaying seaweeds.

Sand dunes can occur inland, but most are found fringing sandy beaches. The onshore winds blow the sand into high ridges, which are then held in place by the growth of a mat of plants such as Marram grass on their surface. When sand dunes are first formed, they have a fairly alkaline soil because of chalky shell fragments mixed with the sand. Rain dissolves the chalk and eventually the dunes become acid and carry heathland.

The smallest soil particles are those of clay and silt. They form extensive mud flats along coasts, especially near the estuaries of slow-flowing rivers filled with mud. These mud flats are covered by the rising tide twice a day, and the different plants growing there are in belts according to the time they are submerged each day. Near the low tide level only the thread-like algae can grow, but higher up the mud more and more different kinds of plants are found. The highest points, out of reach of all but the highest tides, are grasslands.

Cacti and Succulents

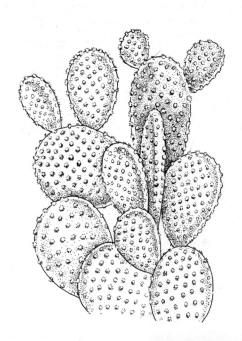
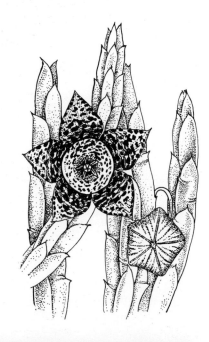

It is one of the many wonders of evolution that plants, which began life in the seas, have species that live in extremely dry areas. These are the succulents. Their thick, fleshy, swollen leaves are filled with water so that they can survive when and where water is not always available. They grow in places where the rainfall is low, unevenly spaced throughout the year and evaporated by high temperatures before it can be used by the plants. Therefore they are mostly plants of deserts, dry grasslands, rocky seaside cliffs, salt marshes and sandy shorelines. Although seaside places are usually wet, often the only water available is sea water. It can be so salty that the plants that live in or near it cannot use it, and so they have developed succulent features.

The most obvious special feature of succulents is their water-filled swollen leaves and stems. Some species, such as most cacti, have lost all their leaves and only the main stem remains. Many succulents live in such hot, dry areas in deserts that their leaves have few water-losing pores. These are sunk in deep grooves that trap a layer of damp air.

Desert cacti grow in the wild only in America, where these large, candelabra-like plants can be seen.

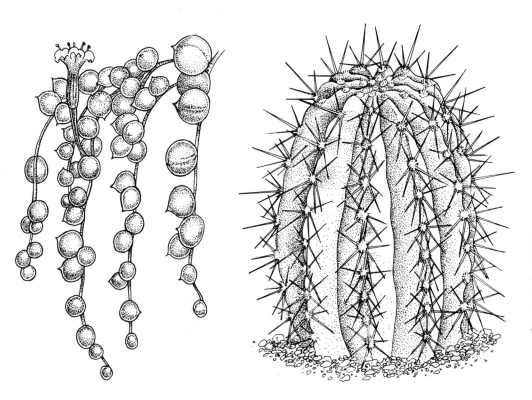

These cacti represent five species: Cotyledon, Opuntia, Stapelia, Ceropegia, *and* Echinocactus. *The first two have succulent leaves, but the other three have water-filled stems on which the leaves are reduced to spines or very small scales.*

Other ways that these plants make sure that water is available include the development of chemicals in the cells that retain water, and an extensive root system. In a typical desert scene the succulent plants are often standing many metres apart. However, below the surface the water-seeking roots of these plants extend over a much wider area. The roots penetrate very deeply in their efforts to find water in underground soil and subsoil layers.

There are succulents in many plant families. Some very large groups only have a few succulent members. For example, the only succulent in the daisy family is a green sausage-like plant from Africa. There are three plant families in which almost every member is a succulent. The best-known is the cactus family, which has 2000 species spread throughout the hot, dry deserts of North, Central and South America. One group, the Prickly Pears, has many species growing all over the world. When they were first introduced into Australia, they spread as a very strong weed and became a threat to national food production. They were finally brought under control by the introduction of an insect that fed on them. Today cacti are found all over the world as garden and house plants. Some have their stems reduced to a spiny ball, which, if it has a long waterless resting period, will produce a mass of orange, crimson or scarlet flowers.

One subtropical family of succulents has fairly large and often brilliantly coloured daisy-like flowers. Some species are grown outdoors in the cooler parts of the world as summer bedding annuals, and a few, such as the Hottentot Fig, have become widespread on sunny cliffs in northern Europe. Although some members of the family are normal in appearance, others are just a pair of very thick and almost round leaves. Some are brown and look like the pebbles and stones among which they live.

Even in the damper and cooler parts of the temperate regions certain habitats can be very dry. Rock surfaces, sand dunes and shingle beaches, stone walls and even buildings and roadsides can get so hot that moisture quickly evaporates. The Stonecrop and Houseleek family has several species that are native to temperate Eurasia and North America. It also includes many tropical species. As well as having colourful flowers, the succulent leaves of these plants can be various shades of green, blue and red, and are often covered with a waxy coating like a ripe plum.

Sedges, Rushes and Reeds

Two groups of flowering plants that resemble grasses in general appearance are the sedges and the rushes. Plants of both families are found throughout the world, but they are most plentiful in the cooler temperate and mountain regions. The great majority of species are found in wet places such as upland moor, marshes, bogs and fens. Another group of broadly similar plants is the reedmace, or false bulrush, family.

As well as forming conspicuous clumps in damp places, and even in running water, reeds, rushes and sedges sometimes form a turf along with grasses. Some species are grown commercially for their stems, which, when dried, can be used for making and decorating furniture, for weaving baskets and for making mats and roof thatching. The stems of papyrus, or paper reed, were used thousands of years ago to make paper. The parts of some sedge plants are edible, such as the Chinese waterchestnut. It is not a true or horse chestnut at all, but the swollen stem of a sedge.

Most plants in these three families are non-woody perennials, but some are annuals. They usually have underground stems from which the upright stems spring. In the rushes the stems and leaves are usually flat and ribbon-like, although in one genus they are cylindrical, long and pointed like large needles. The sedges are unusual among flowering plants because their stems are usually triangular instead of flat or cylindrical.

All sedges, rushes and reeds have small flowers without obvious petals and sepals. The individual flowers are grouped into distinctive clusters.

Rushes and other marsh plants growing in a shallow pool.

A tussock sedge growing at the edge of a pond. Each tussock is a single plant growing on dead remains.

The toadrush grows in damp patches at the edges of tracks and paths, and in the mud near ponds and lakes.

The common reed grows in swamps and estuaries throughout the world, in both temperate and tropical climates.

Grasses, Bamboos and Palms

Coconut palms are found throughout the tropics, where they are grown to provide food, fibres, and oil.

The commonest plants in the world are the grasses. They are also the most important. As well as providing food for most domesticated and wild meat and milk animals, grasses supply much of the world's sugar, in the form of sugar cane, and include all the cereal crops, such as maize, wheat, oats, rye, barley, millet, and rice.

Grasses have fibrous roots. The growth buds either branch at ground level to form clumps, or produce long stems underground or on the surface that eventually give a close turf when intertwined. The upright stems are usually hollow, and can be very tough and fibrous and, in bamboo, quite woody. The leaves grow at more or less regular intervals up the stem, and their bases clasp the stem to form a sheath to protect the delicate growing tissues. Where the sheath meets the leaf blade, there is a small, thin membrane. It is different in each type of grass and is a useful feature for identifying the species.

Grass flowers do not have petals, and individually are much smaller than those of most plants. However, the flowers, called spikelets, are usually grouped to form relatively large spikes, which are up to 2 metres long in some plants.

The leaves of grasses are very similar in appearance and colour. The inflorescences of grasses vary in the size, position, spacing and colour of the individual flowers. These different grasses have developed partly in response to the environment and partly randomly.

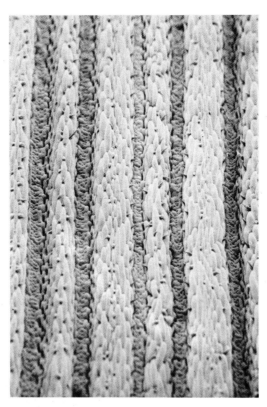

Left, these children in the Solomon Islands are using the hollow bamboo canes to carry water.

Right, this photograph of the surface of a grass leaf was taken with a scanning electron microscope. It shows the minute scales of silica and hard tissue protecting the leaf from insect attack.

The bamboos are a particularly distinct group of grasses. Although they are like grass in structure, bamboos, with their very rigid stems, are often like shrubs or trees in general appearance. They may even grow as tall as trees and some species reach up to 40 metres in height.

Like some trees, many bamboos flower only at irregular intervals, not annually like most plants. In some cases the flowering gap recorded has been over 100 years long, and certain bamboos die after these unusual flowerings.

Grasses grow throughout the world and although bamboos are native only to the hotter areas, they can easily be grown in the cool temperate regions.

Although not technically a tree, the coconut palm is probably the best known tree-like plant in the world. It is related to the grass family and is found throughout the tropical countries. It may grow scattered along the coast or in plantations, where the coconuts are collected, dried and exported all over the world for the manufacture of cooking fats, food, margarine and soap. The flesh of the coconut is a very useful and storable food, and the tough, fibrous husk can be used to make matting. Other palms, such as the date, sago, and oil palm, are almost as well known and are equally important as sources of food for people.

There are 2500 different palms in the world and they grow in all tropical and subtropical areas. They are tolerant of drought and are common trees in desert oases. One species, the dwarf palm, is native to southern Europe and will also grow in sheltered spots throughout the coastal areas of Great Britain and North America. Fibres from its dried leaves are used for making rope and brushes. Another species of hardy palm grows in the mountains of central China, and there are others that are found up to 2500 metres high in the Himalayas.

Palms are unlike other trees in that no true wood is produced in the trunk. The tree's rigidity is maintained by sheaths of strong fibres around the numerous veins running down the trunk. Palms rarely produce branches and all the effort of growing becomes concentrated into a single bud at the tip of the trunk. Old leaves die, and it is their dead bases that form the characteristically scarred bark of a palm tree.

Although the leaves of the various palm trees seem different from each other, they are either fan-like or grow from a single stem and are like huge feathers. The flowers are also varied. Some species bear simple spikes, and others have up to a quarter of a million florets at one time – the effort to produce them is so great that the whole tree then dies!

Fields, Heaths and Moors

One of the first actions of pioneers settling in a previously uninhabited country is to cut down the trees and drain the soils. People have done this throughout Europe and over much of North America, Asia and Australia. The result has been that the forests have been replaced by grassland, which is used for grazing cattle and sheep, or ploughed and sown for harvested crops like cereals, potatoes and sugar beet.

If grazing and ploughing is stopped, coarser herbs will often grow, and in time shrubs and then trees will reappear and eventually a forest will be re-established. An exception to this is where the original forest removal was so extensive that the rainfall pattern has changed, making it impossible for trees to grow again.

Natural grasslands, heaths and moorlands occur where it is either too windy, too dry or too shallow a soil to support woodlands. This means they are found towards the centre of continents away from the rain-bearing winds, on the lower slopes of mountain ranges, and on hard rocks that bear only thin, poor soils. If it is too dry or too hot for grasses and small herbaceous plants, deserts will form. The world's largest areas of natural grasslands are the pampas of South America, the veldt of South Africa, the steppes of central Asia, the prairies of North America, and large tracts in Australia and New Zealand.

The number and kind of plants that grow in both natural and man-made grasslands vary according to the type of soil. Very acid soils are poor in minerals and the number of different grasses that can live in these conditions is small. Chalk and limestone soils, both very alkaline and rich in nourishing minerals, can carry a greater variety of species in a given area than any other plant community in the world. The grasslands that developed on the chalk rocks in England after Neolithic people first removed the beech woodlands and prevented their return by introducing sheep are known to botanists throughout the world for their rich variety of plants. Fewer sheep are grazed there today and these chalk grasslands are maintained by rabbit grazing. On clay and loam soils, especially where they are poorly drained or regularly flooded, grass growth is very rich. These grasslands provide excellent grazing for dairy cattle and hay for winter feeding.

Heaths and moors occur on very acid soils that do not have many minerals. Here grasses cannot grow very vigorously and are replaced by other herbaceous and small woody plants such as heathers. The word heathland is generally used to mean vegetation dominated by heathers, while moorland usually includes heathland, high-altitude poor quality grassland, and intermediate areas where grasses and heathers grow together. Heaths and moors are found over much of the higher ground in Canada and Europe.

Grasslands are maintained by grazing or mowing, but heaths cannot be cut and offer poor grazing. They are maintained by burning, which is carried out every seven to twelve years. Although it may seem to be a drastic method, the plants can survive, and shortly after a fire the heathers and other plants re-establish themselves from their roots. The burning provides a stimulus for seeds hidden in the top layers of the soil to germinate, and within a season most signs of the burning will have disappeared. If heaths are not burned regularly, dead plant matter accumulates on the soil surface. Then when a fire eventually occurs, by accident or deliberately, this layer of dead plants can burn so fiercely that it destroys the soil and the roots of the plants as well.

Where it is too cold or too dry or too windy for trees to grow, fields, heaths and moors will develop. In many parts of the world trees have been cut down, and regular ploughing and grazing keep the land open.

Cereals

Every day millions of people around the world are busy eating grasses in one form or another! Cornflakes, bread, porridge, beer, spaghetti and many other foods are all made from cereals, and cereals are members of the grass family. They contain large quantities of starches, which people need to eat for energy and warmth. All plants contain these foods, but cereals contain very large quantities of them. Because they are easy to grow and produce crops quickly, they provide most of the world's supply of starches.

The seeds of all cereals are packed with starch. Much of the rest of the cereal plant, such as the stems and leaves, has considerable food value as straw for animal fodder and after processing, for human consumption. Cereals can also be fermented to produce alcohol, not as a drink but as an almost pollution-free fuel that can be used to drive motor vehicles! As the supplies of oil are reduced, this use of cereals might become more important. Experiments are even being carried out to use the whole of a cereal plant as a fuel to drive a dynamo to produce electrical energy.

Cereals have been grown for their seeds, or grains, since the Stone Age, but it is only in the last century that they have been scientifically cross-bred to give very high yielding crops. The grains are harvested when ripe and usually have the outer scales removed. Some grains, such as oats and wheat, can then be eaten whole either raw or after cooking. Many cereals, however, are ground up before being used, usually as a flour for baking or as a porridge for boiling in water. Whether as whole grains or when processed, cereals have the great advantage over many other foods of being able to be stored for up to several years without losing their food value.

Rice has been a staple food of the most highly populated areas of the world – China, India and Indochina – for almost 5000 years. Originally from marshy areas, rice has traditionally been grown in the hotter countries in waterlogged soil or deliberately flooded 'paddy' fields. Today new varieties of rice have been produced that can be grown in ordinary fields even in cooler places, and it is becoming an important crop in Australia, southern Europe and much of the United States.

Maize, or sweet corn, originated as a crop plant in tropical Central and South America, and since Christopher Columbus brought it to Europe it has been taken to many parts of the globe. The characteristic tassels that appear early in the season are the elongated female stalks that catch the pollen produced by the male flowers. The corn cob is made of seeds clustered around the remains of the female stalk. The cobs can be eaten raw or cooked when unripe, or allowed to ripen. The ripe seeds are ground to form cornflour, dried seeds can be used for making popcorn, and the rolled and toasted grains are eaten throughout the world as cornflakes. Oil for cooking can also be extracted from maize.

Wheat is the most important food crop in the cooler regions of the world. Vast areas in Europe, the Soviet

People have removed the native plants and replaced them with cultivated cereals to provide food for the world's growing population.

Different cereals are grown in different climates and soils. From left to right, millet is cultivated in the savannah areas of Africa, rye and wheat thrive in cooler regions such as North America and Europe, rice is the staple diet in wetter areas of tropical Asia, and maize is grown in almost every region.

Union, Australia, the United States and Canada are devoted entirely to growing it. Related to wheat and grown for flour and for brewing are rye, oats and barley. Straw, the dead stems and leaves of harvested wheat and its relatives, is used as fodder for farm animals and as a packing and insulating material.

Cultivated almost entirely in the tropics and subtropics, the many species of millets are much better able to withstand drought than any other major crop. They also crop well on poor soils and can be stored for long periods without losing their food value or taste. As well as being eaten as a porridge or made into a flour for human consumption, millets are important throughout the world as food for domesticated fowl.

Plants as Medicines and Drugs

Throughout history plants have been used not only for food, fuel, shelter and clothing but also as remedies for illnesses and as tonics or potions. Modern research has shown that many of the plants used had no curative value, while a few contained very powerful medicines. Many of the 200 000 species of flowering plants as well as a few ferns, conifers and some mosses, liverworts, seaweeds and fungi have been tested for their chemical content, and particular attention has been paid to their medicinal qualities. Although modern technology has enabled many drugs to be made artificially, there are still several plants that are cultivated to produce drugs more cheaply and more efficiently than factories.

Quinine, a cure for malaria, is extracted from the Yellow Bark, a high-altitude tree found in the Andes and now grown in several parts of the world. Atropine, hyoscine and scopolanine are drugs widely used for treating asthma and neuralgia, and for treating eyes. They are derived from the Deadly Nightshade, which grows throughout Europe and Asia. Cocaine, which can be used as an anaesthetic, is derived from the cocaine bush that grows in Peru and Bolivia. Liquorice, used in

Deadly Nightshade is, as its name suggests, a poisonous plant. However, a useful drug – atropine – can be extracted from it, and is used by doctors to control spasms and to enlarge the pupil of the eye for examination.

Foxgloves contain the drug digitalis, which is very important in treating heart diseases.

Colchicine comes from Meadow Saffron plants. Large doses of it given to other plants will double their size and their crops.

The milky sap from the seed pods of opium poppies is dried and processed to make opium, morphine and codeine.

cough medicines and as a laxative, is extracted from the southern European liquorice plant. Eucalyptus oil comes from blue-gum trees, many of which grow in Australia, and is used for clearing catarrh and helping breathing. Castor oil, another laxative, is made by pressing the seeds of a tropical African tree. Morphine, a very powerful pain killer is made from opium poppies.

From earliest time people have eaten or chewed pieces of certain plants, drunk extracts from them, smoked, inhaled and injected them into their blood because of the sensations and moods they induce. If instead of being administered by medical doctors to relieve pain, some of the medicinal drugs extracted from plants, such as cocaine and morphine, are taken indiscriminately, they can become addictive. The user becomes dependent on them and requires increasing amounts, and the result is that the drug can eventually poison and kill him. Some medicinal plant drugs are so powerful, such as those extracted from certain toadstools, that the user may hallucinate and even imagine that he has strange powers, like x-ray vision and the ability to fly – which can have disastrous results!

A native of North America, the tobacco plant is cultivated throughout the world. The chemical nicotine contained in its dried and fermented leaves is a drug with a mildly relaxing effect, but the tarry substances produced with it when it is smoked in a pipe, cigarette or cigar or when it is chewed can be very harmful. Nicotine is also good for killing insects.

Tea, coffee and cocoa contain chemicals similar, although much milder and non-addictive, to those in other drug plants. Cocoa contains the stimulant theobromine and is made from the beans of the Cacao tree. The dried leaves of various evergreen shrubs are used to make tea, and the fermented, dried and roasted beans of coffee bushes are used for coffee, two mild stimulants containing the drug caffeine.

Alcohol is produced by fermenting sugars with a single-celled fungus, yeast. It is used as a drug more than all the other drugs combined, although much of the world production of alcohol is used as an industrial solvent or as a fuel. Cola is also used as a mild stimulant in drinks and when its leaves are chewed.

Insecticides are produced by several plants. They are usually as effective as factory-made chemicals and most have the advantages of not being poisonous to people, other mammals or birds, and not accumulating in the bodies of animals. One of the most widely used plants for this purpose is the Pyrethrum.

Derris is made from the dried and powdered roots of the Derris plant from Malaya. As well as being an insecticide, it is used to kill fish, which can be eaten immediately afterwards without any harmful effects.

Trees

Trees are woody plants that usually have a single trunk and can vary in height from miniature bonsai trees a few centimetres tall to the giants that stretch 80 metres or more towards the sky in tropical jungles. Trees can be conifers, that is, bear cones, or flowering plants. Conifers are plants that bear their seeds in cones rather than in flowers. The male cone produces the pollen, which is spread to the female cone. When this cone is ripe, it opens and scatters the seeds it has produced. Most of the conifers are evergreen but the flowering plants can be evergreen or deciduous, losing their leaves at a set season.

Trees live very much longer than non-woody plants. The trunk, branches and twigs usually increase in length each year, and many of the smaller, older and weaker stems are shed. Some stems stop growing in length after reaching a certain size, but they increase in diameter. All woody parts of trees increase in diameter as layers of new growth are added both towards the centre and the outside of the stem. The tissue that produces this new growth is called the cambium. Chemical changes occur in the inner part of the stem to form wood. This inner wood growth presses against the outer layers of the stem, which become strained and eventually crack. To replace this damaged layer there is another cambium tissue, called the cork-cambium. It produces an outer layer of elastic cork. In some species the cork-cambium can continue to produce cork layers that dry and continually are lost, thus giving a smooth new surface to the tree. In many species, however, the original cork-cambium dies, but is not shed. It is replaced by progressively deeper cambiums. In these cases the stem surface has the familiar rough and often cracked or flakey, scaley appearance. This series of dead and alive tissues is called the bark, and protects woody plants, such as trees, from browsing animals, insect and fungus attack, and extreme conditions.

Sap is either the contents of the large, watery cells in plants or the liquid contained in the veins. It can be extracted from trees, or tapped, by making a slit in the trunk and piercing the veins. The sap from the Rubber Tree is used to make rubber, and maple syrup is made from the sap of maple trees, particularly the Sugar Maple.

Trees are widely cultivated for timber production, to shelter agricultural land, to screen unsightly buildings

Trees may provide food and timber for people and a home for animals.

and for decoration. In every country there are both native species and those that have been introduced from somewhere else. To provide trees for a greater range of habitats, and to improve features such as timber production, hardiness, and disease-resistance, selected plants are cross-bred, or hybridized.

Before people developed agriculture, most of Europe was covered by forests. In northern Europe and on the lower slopes of mountains the trees were mainly evergreen conifers with an intermingling of deciduous species such as birches. In contrast, in the southern part and in the western coastal zone the forests were composed mainly of deciduous trees with just a few conifers, such as pines, junipers and yews, in unusual habitats. Today over 75 per

Trees always have a single trunk, but the side branches are of various sizes and shapes so that each tree species has its own characteristic profile. In most flowering trees the branches usually point outwards and upwards, while most conifers, like the redwood and larch, have drooping branches that prevent a build-up of heavy snow.

Far left, the redwood. From left to right, top row, hickory, oak, larch and poplar; bottom row, teak, sapele and Eucalyptus.

cent of this cover has been destroyed to provide land for farming, building and transport, or replaced by densely packed timber plantations, usually of faster growing alien species. The remaining oak, ash, beech, birch and conifer woodlands are still a very characteristic part of the European landscape and are very important in maintaining the diversity of wildlife, maintaining and controlling the water vapour in the air, and in preventing soil erosion by wind.

Much of inland North America has always been so dry that woodlands never occurred there – the original prairies and deserts were quite natural. Nevertheless, the total area of woodland is very much greater than that of Europe. It varies from conifer forests in the sub-arctic areas of Canada and on mountain slopes everywhere to an evergreen subtropical jungle in Florida. Many of these are visually spectacular trees, such as the giant Californian redwoods, and the colourful hickories, maples, poplars and birches. In Canada alone there are over 150 different native trees, and more than four times that number in the United States.

Equatorial lands, except where they are very dry or at very high altitudes, are the home of jungle trees, called tropical rain forest. Trees of the wet tropics do not have to possess any special features to cope with an unfavourable season. They grow at more or less the same rate every day of the year, can flower and fruit at almost any time, and shed their leaves throughout the year so that they remain evergreen. Because of the high rainfall the leaves often develop pointed tips to conduct the water away. Many tropical trees are 100 metres tall and develop buttress or stilt-like roots to improve their stability.

Australia has a very wide range of vegetation, varying from tropical jungle to desert, and includes many types of trees. The southern beeches and the blue-gums, or Eucalyptus, are widespread in both wet and dry areas, and many are widely planted for timber. Eucalyptus trees can be as much as 100 metres high and have a trunk 8 metres round. It is their waxy, often blue-green leaves and their lidded flower buds that make them different from any other trees. Other Australian trees, such as the Australian honeysuckles, the aptly-named bottle-brushes, and the wattles are famous for their brilliantly coloured and unusually shaped flowers.

Shrubs and Climbers

Like trees, shrubs are woody, perennial plants. Generally, however, they do not have a single trunk, but many branches growing from the base, and they are not as tall. It is possible by careful pruning to turn many trees into shrubs and many shrubs into trees. Even in the wild many trees can become shrub-like because of bad growing conditions, such as shallow soil and high winds. Shrubs occur throughout the world, either as an understorey in woodlands or as the main feature of the landscape. They grow at the margins of woodlands and extend the area of woody plant growth far into the desert or grassland.

Shrubs are widely grown in all countries, both as garden plants for decoration and for crops. They provide many different crops, from fibres such as cottons and drugs such as cocaine to nuts and many soft fruits, such as raspberries and gooseberries. Only rarely is shrub wood used as timber for building or furniture, but several shrubs are grown for their long, straight stems. These are used as garden stakes and bean poles, and, in Europe, those of hazel are used for making an ancient form of fencing called hurdles.

The Spanish or sweet chestnut is technically a tree, but it can be periodically cut down to ground level to encourage the growth of many miniature trunks. These are harvested every seven to ten years. Then, after being split lengthways, they are used to make split chestnut fencing, which is used in parks and gardens everywhere. Chestnut is a very durable wood and frequently lasts much longer than the iron wire used to join it together!

Many flowering plants, both woody and non-woody, have weak stems that require the support of another plant, rock or wall. Although they usually climb, twine or scramble upwards towards the light, they can also creep along the ground. Sometimes these plants have clinging hooked thorns, clasping roots, or tendrils to help them climb. The force exerted by some twining plants, such as the tropical figs, can be so great that the tree that the plant is climbing on is strangled and eventually dies.

Many climbing and scrambling plants are parasitic, obtaining all or some of their food from the living tissues of the host to which they cling. Some tropical parasitic climbers grow faster than any other plant.

Shrubs have no main trunk, but many stems arising at or just above soil level. Mature shrubs, such as the Berberis, Skimmia, Hydrangea *and* Rhododendron *shown here, are rarely as large as full-grown trees. They are therefore very useful as garden plants, although they have to be pruned to prevent them spreading out and smothering other garden plants. Climbers, like the* Clematis *on the right, are often grown to hide old tree stumps and unsightly walls and fences.*

Despite being too weak to stand upright without support, the stems of climbers are often very tough. They are used as string and rope by craftsmen in some of the non-industrialized countries.

Climbing plants have long been favourites with gardeners because they can be used to cover unsightly objects such as dead tree-trunks and old garden sheds. Towns and cities can be made much brighter by having the fronts of many houses and walls covered with climbers such as vines and virginia creepers, whose large leaves turn brilliant red, brown, or orange in the autumn.

Creeping plants, usually called ground cover plants, are used in gardens and orchards to cover otherwise bare ground and to smother weed growth. Creepers are planted on a larger scale to control soil erosion by, helping to hold in place banks of loose soil and rock.

Plants can climb by twining but are usually helped by clinging hooks, left, suction discs, centre, or clasping tendrils, right.

Forests and Woodlands

In forests and woodlands the main type of plant is trees, and they form a continuous cover to the ground. Woodland usually refers to smaller areas of this type in the cooler parts of the world, while forest is often used to mean large tracts of jungle in the tropics. In the Middle Ages forests in Europe were almost any large area of tree growth, often interrupted by areas of heath and grassland. Today, larger woodlands and heaths are frequently called forests such as the New Forest in England, the Black Forest in Germany, and the Redwood Forest in North America.

Woodlands are much more complex than most other plant communities, such as grasslands or moorlands, mainly because they are made of three or four layers of plants. The most important layer is the tree layer, which is composed of the larger, more vigorous and more mature trees. Underneath are the younger and smaller trees and shrubs that form the shrub layer. If the shade cast by the trees and shrubs together is not too dense, there will be enough light for a layer of early flowering herbaceous plants under the shrubs. On the soil surface, but under the herbaceous layer, there can be a covering of mosses, liverworts and lichens as well as seedlings of herbaceous plants, shrubs and trees that make up the ground layer.

In tropical rain forests there can be three or more tree layers, which cast such a dense shade that often the shrub layer, herbaceous layer and ground layer are missing. Even in cooler areas it can be so very dark under trees such as beech and pine that the only plants that will grow are those kinds that do not require much light and obtain their food from dead plants and animals, or parasitically from the roots of the living trees.

The forest trees of the world are of two distinct kinds. On either side of the equator in the damp, tropical regions and in the much colder areas towards the polar ice-caps the trees are evergreens. In the temperate areas between these regions, in a great wide belt across Europe, northern Asia and North America, the trees are mostly deciduous and lose their leaves in winter. Trees in the tropical forests and in the temperate area are mainly flowering plants. Those in the colder regions are usually conifers.

The plants in forests and woodlands grow in layers. The mature trees form the highest layer. Below them are the younger trees and shrubs, and the herbaceous plants and tree and shrub seedlings. The soil is covered with a layer of mosses, liverworts, lichens and fungi.

The woodlands of the temperate parts of the world formed much of the vegetation after the Ice Age. Since then they have been cut down to provide space for agriculture and building. Today many woodlands are protected and managed to produce a continual supply of timber and to maintain the great variety of plants and animals found there.

There is always plenty of warmth and moisture in tropical jungles, and the trees, climbers and herbs grow very rapidly. Flowering and fruiting occur at all seasons and young plants are found everywhere.

The evergreen forests do not vary greatly in appearance throughout the year, although each species tends to flower and fruit at certain times. In the deciduous woodlands there is a marked change in the appearance from one season to another, as the trees lose their leaves and regain them several months later. This tree leaf-fall and growth influence the light and moisture conditions under which the shrub, herbaceous and ground layers exist, and so the layers have very definite and distinct flowering and fruiting times. Thus deciduous woodlands can have quite different plants growing at different seasons as well as the trees themselves looking quite different.

Woodlands and forests, with their several layers of plants and often changing seasonal appearance, provide a very wide variety of homes for animals, especially winged kinds such as insects and birds. The deciduous woodlands of Europe and North America contain a greater range of plants and animals than any other temperate plant community, and the tropical rain forests have been estimated to hold almost half of all the different animals and plants in the world.

Herbaceous Plants

Any flowering plant that has no persistent woody parts is technically a herbaceous plant. Therefore this category includes not only the ordinary plants of gardens and the countryside but also cereals, grasses, sedges, rushes, water plants, weeds and all bulbous and tuberous species. Herbaceous plants can be annuals, which means that they complete their life cycle from seed back to seed again in a single year, biennials, which take two seasons, or perennials, which can last for several seasons.

To the gardener, herbaceous plants are those that die down to the ground each winter and sprout into life again when the growing season comes around. A very conspicuous feature of many gardens is a long, narrow flowerbed filled with herbaceous plants, which provide a wealth of colour throughout the summer. These herbaceous borders may contain a few woody or bulbous plants, but are mainly used for growing a mass of closely packed herbaceous plants.

Most of the world's herbaceous plants originate in the tropics and cannot be grown successfully outdoors throughout the year in Europe and North America, but many of them are ideal for growing in greenhouses and conservatories. Among favourite tropical herbaceous plants are orchids, and bromeliads, which are members of the pineapple family and are watered through the cup formed by the leaves at the top of the plant. Several herbaceous plants are cultivated not for their flowers but for their leaves, including the begonias, and tropical vines and ferns.

The name herbaceous plant is often shortened to herb, but to most people herbs are plants, such as mint, thyme and sage, grown for the scents and flavours that their leaves, roots and flowers give to food and drink. Herbs have been grown for their real and supposed medicinal value since the beginning of civilization, and are still used by many people instead of chemically produced drugs. Some herbs are used in the preparation of cosmetics and pesticides, and as dyes.

Orchids are the largest family of herbaceous plants and are found in every country of the world. The Ladies' Slipper Orchid grows in Malaya and Thailand. Equally spectacular flowers are found on related species in Europe and North America.

One of the largest groups of herbs are the mints. There are many different flavours of mints and cross-breeding has produced many more. The most famous mint is peppermint, which produces the oil used as a flavouring in many sweets, cordials and medicines. Very similar to peppermint is spearmint, which is widely used in chewing gum and toothpaste, and for making mint sauce. Other herbs commonly grown for the flavours and scents of their leaves are sage, thyme, and rosemary. Fennel and dill are grown for their flavoured leaves and seeds, and ginger and liquorice for their roots.

The cereals and most of the root crops are technically herbaceous, as are the salad plants, lettuce, chicory, endive and cress. The many varieties of the herbaceous *Brassica oleracea* include cabbages, kale, brussels sprouts, cauliflower, broccoli and kohlrabi, which are grown in the cool temperate regions. Most cabbages are green, but there are also red-leaved ones, and ornamental cabbages with bright pink or yellow inner leaves.

Orchids are found in every country and in nearly every type of countryside from the tundra wastes of Greenland to the hot, humid rain forests of Borneo and New Guinea. Millions of orchids are grown for the colourful and attractive appearance of their flowers. These are unusual in that one of the three petals is transformed into a very noticeable and spectacular lip, which attracts pollinating insects and acts as a landing platform for them. The only useful orchid is *Vanilla planifolia*, whose fermented seed pods produce the vanilla flavouring used in so many foods and sweets.

Garden chrysanthemums, which are members of the daisy family, are grown all over the cooler parts of the world and there are native species in many countries. The large mop-headed flowers seen in shops and in nurseries are hybrids, the first of which came from China and Japan hundreds of years ago. The flowers on these plants are really masses of little flowers, or florets, grouped together very closely.

In many gardens and parks plants with a wide range of flower colours are carefully chosen to provide a show throughout the spring and summer.

Herbaceous plants are not grown just for their flowers – most of our food plants are herbaceous. Cabbages are usually biennial.

New Plants from Old

Plants have always been a major part of people's food. At first seeds from plants collected in the wild as food and probably dropped on the ground germinated and grew into mature plants around nomadic camps. By 6000 BC in Palestine people had become settled and grew crops rather than searching for suitable plants in the wild. About 1500 years later people not only grew many crops from seed, but there is evidence that they selected and used seeds only from the strongest, most productive plants.

This process of selection by using only the best plants is still the most widespread method of producing new plants from old. Today plants are selected not only for yield, but also for such features as disease resistance, the ability to withstand bad weather and poor soil, and speed of ripening.

Selection to produce the required new plant is usually a very long process, and can involve the equivalent of up to twenty years of growing seasons. Occasionally, however, a quite different individual appears spontaneously in a crop. This is called a sport or mutation, and it is caused by a sudden change in the way that particular specimen is made. Many of our brightly coloured garden flowers are sports that have arisen in cultivation. Among food plants, red and green cabbages, broccoli, cauliflowers, and sprouts all began as sports from the wild cabbage.

During the last sixty years new varieties of plants have been raised by artificial cross-breeding, or hybridization. The yield and the quality of crop plants, and the range of flower colours of decorative plants have been increased by scientific breeding carried out by government research stations, universities and private seed firms and nurseries. Hybridization has enabled breeders to produce particular plants for particular conditions and needs. Potatoes that are very resistant to disease, grow well on poor soil and are useful for mashing, boiling, frying and roasting have been produced in this way. To achieve such results, breeders select an existing strain that has some of the desired features and pollinate it with another strain that has other desired features. To reinforce certain desirable characteristics the hybrid is then

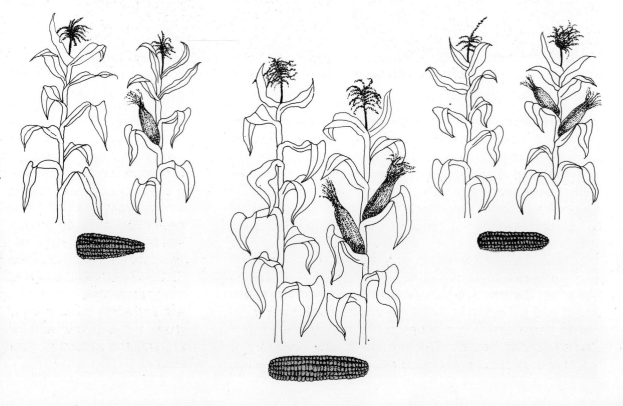

Plants are carefully selected for such qualities as rate of growth, resistance to disease, draught and frost, as well as value and size of crop. They are then cross-bred to produce new strains, which are of enormous importance in feeding the world's growing population. The vigorous maize in the centre was produced from the two parents shown on either side.

The original wild cabbage had a few leaves that could be eaten. Careful breeding has given us a variety of useful foods. Above are a green cabbage, a red cabbage sliced open to show the arrangement of leaves and stem, and brussels sprouts. Left is a cauliflower and next to it is sprouting broccoli.

pollinated with one of its parents or with a similar hybrid, and this process is repeated until the characteristic is much more obvious than in either parent.

In the past these breeding programmes could take many years, but today they can be greatly shortened by giving the plants involved an abnormally short life cycle by controlling growing conditions such as day length by artificial lights.

In 1943 a plant breeding programme was started in North America to produce high-yield wheat for Mexico and other Central American countries. The hybrids produced had the added advantage of maturing quickly, so that in some areas two crops could be harvested in a single growing season. The success of this breeding of wheat led to similar programmes with rice and maize.

This 'Green Revolution', as it has been called, has had a very beneficial effect on food production.

Unfortunately, these high-yield strains can produce new problems, such as soil destruction, because of the large quantities of chemical fertilizer needed to sustain the high yields. New pests and diseases that are not affected by pesticides have appeared. In cross-breeding for high yields it is not always possible to incorporate disease resistance in the same strain, and entire crops of some 'Green Revolution' plants have been destroyed.

Scientists are now turning their attention to selecting and breeding other sorts of plants that could be used for food or fuel. Algae and fungi, grown in tanks of diluted sewage and industrial waste, may well be the food of the future.

City and Roadside Plants

In order to grow crops and build houses, factories, roads, railways and airports people have destroyed many of the world's wild places. It has been estimated that as much as a third of the world's 300 000 species of wild plants are threatened with extinction.

However, human activities do not inevitably result in the destruction and loss of plants. Fields in which crops are grown can contain many weeds that would not have been able to grow in a natural area because of competition. The walls, roofs and gutters of houses, factories, churches and railway stations can provide very good and well-protected places for plant growth. The heat used to heat buildings leaks through the walls and can make the temperature in a city 2°C higher than that of the surrounding countryside, and this permits many foreign plants to grow.

Derelict building sites, car parks, rubbish tips and areas left over after construction form ideal places for plants. Infrequently cultivated and poorly maintained public parks, private gardens and allotments, cemeteries, churchyards, factory areas and even the centres of traffic islands, roadside and railway verges, cuttings and embankments contain many species found nowhere else in a country. Many of our common garden weeds have spread solely by the accidental transfer of seeds when plants are bought from a nursery.

Many weeds are wildflowers from the surrounding countryside that find the conditions in a garden much more suitable than in the wild where they have to compete with many other plants. Herbaceous St John's Worts, blackberries, and dandelions are examples of weeds that have spread in gardens, agricultural and forestry areas all over the world. Willow-herbs from Europe, North America and New Zealand are found as weeds in all of the cooler parts of the world.

Weeds are usually defined as any plants that are growing where they are not required. Therefore a weed in one part of a garden could be a crop or decorative plant elsewhere. To botanists and farmers weeds are useless plants that are a nuisance because they compete with useful plants for water, minerals, light and space. If they are not removed or controlled, they can overpower the cultivated plant and reduce its yield. Some weeds can produce substances from their roots that slow down the rate of growth of neighbouring plants. Weeds can also have diseases and pests that spread to their cultivated neighbours.

Weeds are very quick growing plants that spread by producing large quantities of rapidly germinating seeds and have roots that creep outwards from the plant to give rise to new plants in any available space. Some of the most difficult weeds to control are those, like bindweed and couch grass, that have fragile roots that break into small pieces when the soil is disturbed. Each piece of root develops into a complete plant.

Some weeds appear only when growing conditions are ideal for cultivated plants. There are several species that will flower, become fertilized, shed and germinate seeds at all seasons of the year and are not affected by very low temperatures and snow. A good example is the Annual Meadow Grass, which occurs in many countries.

Many weeds in farming are foreign in origin, and were introduced with crop seeds from another country. In their original countries they could have been kept under control by their own pests and diseases, but when they are introduced elsewhere they grow vigorously because there often are no natural enemies present. Weeds can be controlled by introducing insects and diseases, but care must be taken to ensure that these do not attack the crop plant as well.

Many plants that are rare in the wild will grow very well in man-made surroundings. As far as the plant is concerned a brick wall or stone pavement is the same as an outcrop of natural rock.

Although motorways take up a lot of land that could be used for growing crops, their banks form very good nature reserves on which people rarely trespass. Pollution from vehicles means that very few plants survive next to the road itself, but further away a great variety grows.

Growing Plants Indoors

You do not need a laboratory or garden to study botany. You do not even have to go to the countryside or into a public park. The forms of fruits and vegetables, the structure of the leaves in a vase, the way a plant reacts to light, and the nature of flowers can all be examined indoors.

Plants can be grown very easily indoors for their decorative features, such as leaf texture and flower colour, for their scents, and for their food value. Home winemakers and mushroom growers are just as much plant growers as people who specialize in cacti, ferns, begonias or other fashionable houseplants. In every case success is possible only when all the conditions for the plant growth have been met. A plant can complete its life cycle from seed to seed only if it has the right amount of light, heat, water, air and minerals at the right time and in the correct order.

Houseplants for decoration are usually grown for their leaves or their flowers. The dry atmosphere to which plants are subjected in an ordinary living room is often not suitable for a plant to complete its life cycle and it will remain in the non-flowering state, especially if there also is little light, which is ideal for foliage plants. If the plant is grown for its flowers, it is usually necessary for it either to be bought in flower or put outdoors or in a greenhouse if it is to continue to flower. Bulbous plants, such as crocuses, narcissi and tulips, for indoor use are commercially treated before they are sold by being stored in controlled cold conditions to speed up the production of flower buds.

It is very easy to grow many plants indoors for just part of their life cycle, and more and more people are using germinating seeds for food. The salad plants mustard and cress can be grown very simply by scattering the seeds on a piece of damp blotting paper in a saucer. If the blotting paper is kept damp and the temperature does not fall below 10°C, the seeds will germinate in about four days and the minute seedlings, with just their two green seed-leaves, will be ready to eat in another five or six days. Bean sprouts germinate at about 15°C. These plants do not need much light, as the seedlings are eaten before the chlorophyll has developed.

Indoor plants are often grown only for their leaves. To enable them to flower would mean increasing the amount and strength of light to levels that would be uncomfortable for people living in the same room.

Bonsai trees and miniature greenhouses, or bottle gardens, making interesting and unusual houseplants.

Bulbs and corms planted in a bowl will flower indoors just as well as in a garden.

To understand more about seed germination large seeds can be germinated indoors and the results of different conditions on them can be noted. The best seeds to use for this are beans such as the broad bean, the runner bean, and the garden pea. Decide which kind of beans to use. Then put a piece of crumpled-up newspaper into each of six jam jars. Place one between the paper and the glass in each jar, so that it can be seen at all stages. In one jar keep the newspaper damp and put the jar in the refrigerator. In the second jar keep the newspaper damp and put the jar in the kitchen. Keep the newspaper in the third jar dry and put the jar in the refrigerator. In the next jar keep the newspaper dry and put the jar in the kitchen. Fill the fifth jar with water and put it in the refrigerator, and fill the last jar with water and keep it in the kitchen.

After about a week examine the seeds in the jars. You will see that any of those kept in the refrigerator, in dry conditions or under water (which excludes the air) will not germinate. To germinate most seeds require warmth, moisture and air.

Growing Plants Outdoors

Gardening is a major hobby in many parts of the world. Not only are plants grown for food and decoration or as miniature versions of the countryside, such as lawns and rockeries, but the exercise in digging, hoeing and grass-cutting is very important as a way to keep fit.

The plants that you can grow in a garden depend firstly on the climate, especially the temperature and rainfall, secondly on the kind of soil, and thirdly on how much time and effort you can spend looking after them. In many gardens plants need protection from wild and domesticated animals, birds, and other pests and diseases. It is also usually necessary to control the space available between plants so that one kind does not kill its neighbour by growing over it and depriving it of water, light or air. Vigorous weeds have to be removed, as they compete with garden plants and can carry pests and diseases.

To grow as wide a variety of plants as possible it is necessary also to make sure the soil remains suitable. This means adding nutrients such as manure, compost or chemical fertilizers regularly, and watering the ground if there is not enough rainfall to keep the soil moist. However, it is just as important to prevent the soil from becoming waterlogged by over-watering as it is to prevent it from drying out completely.

Many of the plants grown in gardens are not native to the country in which they are grown. Although some species will grow, flower and fruit in almost any area of the world, many garden plants will only complete their life

A small garden can provide sufficient fruit and vegetables for a family of three. Some will be eaten as they are harvested, but many plants, such as potatoes and marrows, can be stored until required.

A box filled with soil and fastened to the window can be used as a miniature garden. As long as the soil is not allowed to dry out, flowers, and even some fruits and vegetables, will thrive in this unusual situation.

cycle, from seed to seed, in their native areas. If this did not happen our gardens and the countryside would be overrun by introduced species, and many original native plants could become extinct. As it is, many of our garden weeds were first introduced from overseas as decorative plants but found the conditions favourable enough to become naturalized.

Even if you have no garden, you can still grow plants outdoors. Boxes or pots of soil on a window sill or in the smallest concrete patio will enable you to grow a surprisingly large variety of flowers and vegetables, and even fruit such as tomatoes and strawberries.

One of the most successful kinds of plants to grow in pots are bulbs such as crocuses, daffodils, hyacinths and tulips. To prolong the flowering season two, three or even four layers of bulbs, separated by soil, can be planted in one pot. Those nearest the surface will flower first and those deepest in the soil will flower last. After the bulbs have flowered, the leaves will soon shrivel away and other plants can be put in. If the soil containers are large enough and at least a half metre deep, you can grow a small shrub or tree in them, and the bulbs and bedding plants can be planted around its base.

Pots and boxes can also be used to grow climbers such as scrambling or climbing roses, ivies, or an annual fast-growing climber such as the nasturtium or Morning Glory. These plants can climb up taller plants in the pot, sticks, or any structures such as a fence, wall or drainpipe.

In many cities where there are few public parks or private gardens people have transformed small backyards into gardens full of interesting and attractive plants, grown in boxes or pots, or in hanging baskets where there is very little space.

Gardens can be very expensive to stock if you buy all your plants from a nursery or garden centre. In addition, the shock of putting plants in a new situation in your garden when they may have spent the early part of their lives in the luxury of a nursery can kill them or stunt their growth. However, nearly all garden plants can be grown from seeds, which are not very expensive to buy. Plants raised from seeds are usually grown more successfully because they can get used to their environment from the time they germinate.

In some places it is not possible to have soil containers and the soil itself, if it has to be bought, can be expensive. This problem can be overcome by putting stones, sand, or even granulated or shredded plastic into a container and planting into it. These materials will anchor the plant and food can be given by watering with liquid manure. This soilless method of growing plants is called hydroponics. It can result in very healthy and vigorous specimens, although very large plants may become top-heavy and require more water than it is possible to provide on a hot day.

Extinction and Conservation

Plants that cannot adapt and evolve to survive changing conditions become extinct. Since plants first occurred thousands of species have disappeared because they were unable to change. However, the extinction rate has never been as fast as the rate of evolution of new species. The result is that there is a greater diversity of plants in the world now than at any time in the past.

All living things – both animals and plants – grow in ecosystems, which are the plants and animals themselves, the water, air, soil and rocks, and the relationships between all of these. Ideally, all of these things are delicately balanced in an ecosystem, with no single part getting out of hand and affecting the others. Obviously, sudden floods, volcanic eruptions and earthquakes, or the appearance of ice-sheets can upset the balance. It is soon restored, often with some of the original species absent. New species that evolve can also disrupt an ecosystem when they first enter it, but before long they either become extinct or contribute towards the balance. However, one species has appeared on Earth that upsets the balance of any ecosystem it enters and can seriously reduce that ecosystem's natural diversity. We are referring, of course, to people, or, to be more precise, to modern, industrialized, technological and resource-consuming people.

For the first 60 000 years that people lived on Earth they were part of the ecosystem, but about 9000 years ago they started to live in settled communities. They grew crops and domesticated animals that grazed on land that originally was natural vegetation. The food supply was no longer the only major factor controlling their reproduction and survival. They became more and more abundant and destroyed more and more plants to provide space for their own activities. By 1825 there were 1000 million human beings, by 1925 there were 2000 million and in 1975 there were nearly 5000 million. By 1999 there will be 6500 million people on our planet, and the effect on the plants of the world could be disastrous.

To feed, clothe, shelter, educate and entertain this large population, enormous quantities of natural materials and space are needed in addition to cultivated plants and domesticated animals. To obtain this and to process it for human use, many plants have been made extinct, and thousands more are now so rare that they will soon disappear. People's destructive activities affecting plants include the draining of marshes, bogs and fens, the removal of trees and hedges – which encourages soil erosion – peat removal, careless and excessive use of weedkillers, the over-collecting of decorative and drug plants, the pollution of air and water, and the introduction of vigorous artificially bred plants that overwhelm natural species.

As our destructive activities have increased, so has our awareness of them. Appalled and worried by our actions, we are now turning towards protecting and conserving the environment. Legislation preventing the uprooting of rare and threatened plants, or their unlicensed import or export exists in many countries. Laws controlling building development, pollution and land use also help to conserve plants.

Extensive educational campaigns on conservation are proving very successful, and many official and voluntary organizations have been set up locally, nationally, and internationally to conserve plants and animals.

Plants can be conserved in different ways depending on the life cycle, form and habitat of the threatened species, and on the nature of the threat. Ideally, a plant should be conserved in the wild, but this is not possible when the site in which it grows is going to be destroyed. In this case the plant must be taken into a garden or greenhouse where it can grow and reproduce.

Many plants are conserved by being kept in the wild in specially protected and carefully managed nature reserves or plant sanctuaries. In these areas all threats to the rare plant are controlled and the plant is encouraged to make seeds so that it can spread.

As a safeguard against a plant becoming totally extinct because its last home could not be protected or because it was too late to transplant it into a safe place, seeds of rare plants are being collected and stored in seed banks.

Plants may be destroyed by human action, such as the cutting down of forests to build roads or provide land for growing crops. They can also be killed by natural disasters, such as sudden floods.

We can protect and conserve our environment. Marram grass can be planted to restore sand dunes trampled by holidaymakers, and some diseased trees can be saved by injecting them with special chemicals.

Under carefully controlled conditions of humidity, temperature and pressure, seeds of most plants retain their ability to germinate almost indefinitely and certainly much longer than if they were stored in ordinary conditions.

All uncommon species of plants are being carefully observed over the years to see if they become rarer. In some cases expeditions have to be sent to areas with rare and unusual plants to report on their numbers and to try to find out why they are rare. When this information has been studied, it is then possible to suggest what kind of conservation measures to take.

Scientists are also studying other environments in which plants can be grown. They are even testing the effects of weightlessness on plants as part of the space exploration programmes. Simple plants such as bacteria and moulds have been carried in spacecraft and have not been affected by the lack of gravity or by the enormous pull of the craft as it accelerates. These plants are much more responsive to changes in the quantities and quality of food available than to speed or gravity. Experiments with flowering plants have also shown that the light, temperature and the composition of the air around them are still the most important factors, just as they are on Earth.

Glossary

Alga: (plural *algae*) small flowerless plants, often single-celled and able to swim

Annual: a plant that completes its life cycle in a single year or less

Bacteria: microscopically small fungi that often cause diseases

Bark: the tissue of dead cells covering and protecting the stems of woody plants, such as trees

Biennial: a plant that completes its life cycle in two seasons

Bonsai: a tree that has been dwarfed by pruning the roots and growing it in a restricted place such as a flower pot

Bud: a growing point on a plant that may produce a branch or flower, or remain in a resting stage to enable the plant to survive an unfavourable season

Bulb: a underground part of the plant that stores food in its leaves so that the plant can survive an unfavourable season, and reproduces the plant in the growing season

Cactus: (plural *cacti*) a succulent plant growing in hot, dry conditions that stores water in its swollen stem and has spikes for leaves

Catkin: a spike of small flowers dangling and waving in the wind to release or capture pollen

Cellulose: a carbohydrate that strengthens a cell wall and forms fibres in stems

Chlorophyll: a green colouring matter that is found in all green plants and is needed for photosynthesis

Climber: a weak-stemmed plant that clings to other plants or objects and grows upwards to reach the light

Conifer: a tree that has cones instead of flowers

Corm: a short underground stem that stores food and reproduces the plant

Deciduous: a plant that sheds its leaves at a particular time of year and remains leafless until they grow at another time of the year

Dormant: a resting stage, which may refer to a whole plant during the non-growing season or to a bud that only grows if the main bud is destroyed

Evergreen: a plant that keeps its leaves for more than a year, not shedding them at any particular time

Fern: a flowerless plant that has leaves with veins

Flower: the part of a flowering plant that produces the pollen and then the seeds for reproducing the plant

Fruit: a part of a plant that contains a seed or seeds, and develops after the flower has been fertilized

Fungus: (plural *fungi*) a simple plant that cannot make its own food, but takes it from dead or living plants or animals

Grass: a flowering plant with ribbon-like leaves and with no petals or sepals

Herb: an herbaceous plant, especially one that produces scented or flavoured leaves that can be used in cooking or medicine

Herbaceous: a non-woody flowering plant

Horsetails: a group of flowerless plants related to ferns

Hybrid: the result of cross-breeding two plants that are not alike

Hyphae: the individual, usually thread-like, parts of the body of a fungus

Inflorescence: a group of flowers on a single main stem

Leaf: the main food-manufacturing part of most flowering plants

Lichen: plants that consist of an alga and a fungus growing together for mutual benefit

Liverwort: a flowerless plant without veins that reproduces by spores

Meristem: the region of actively dividing cells found at the tip of a stem or root

Moss: a flowerless plant without veins that reproduces by spores

Mould: any small fungus plant that forms a powdery or skin-like covering to its food

Mushroom: a name given to any fungus that produces a large, definite-shaped reproductive body above ground and usually can be eaten

Nectar: a sugary liquid produced by many flowers, usually by the petals, which attracts pollinating insects

Node: the region of a stem where a leaf or secondary branch grows

Nut: a hard, dry fruit that contains only one seed and does not split open when ripe

Palm: a tree-like plant related to the grasses

Parasite: a plant that lives on or in another living plant or living animal and takes food from it, but gives no benefit to its host in return

Perennial: a plant that grows for more than two years, dying only when old or killed

Petal: an individual part of a flower, usually differently coloured from the rest of the plant

Photosynthesis: the process by which plants convert energy from the Sun with the aid of chlorophyll, oxygen and water to produce sugars; the source of all food in the world

Pollen: the powder-like male sex cells of a flowering plant

Respiration: the process by which the plant converts stored foods into energy

Rhizome: an underground creeping stem

Root: the part of a plant that usually is underground and absorbs water and minerals from the soil and acts as an anchor

Sap: the watery content of cells and the contents of the veins

Saprophyte: a plant that lives in or on a dead plant or dead animal and takes food from it

Seed: the part of a fertilized flower containing the embryo and stored food, and which is usually covered by a tough protective coat

Sepal: an individual part of a flower, usually green and often shed when the flower opens

Shrub: a woody plant with more than one main stem

Spore: an asexual reproductive body produced by all plants except seed-bearing plants, which germinates to produce sex cells or a new plant

Stem: the part of a plant carrying the leaves, reproductive organs and roots

Stigma: part of the female stalk of a flower that catches the pollen

Stoma: (plural *stomata*) a pore on a leaf that opens and closes to allow air and water vapour to pass in and out of the leaf

Succulent: a fleshy plant or part of a plant that stores water

Symbiosis: different species living together, each providing something the other cannot provide for itself

Tendril: a modified stem or leaf used for climbing and clinging

Toadstool: a name given to any fungus that produces a large, definite-shaped reproductive body above ground, and usually is inedible or poisonous

Transpiration: the flow of water from the root of a plant, through the stems into the leaves and out into the air

Tuber: an underground stem or root swollen with food reserves

Index

algae 6, 8, 17, 20, 37, 42–7, 52–3, 54, 58, 63, 85, 94
annuals 32, 36, 66, 82, 94

bacteria 8, 16, 38–92, 40, 41, 47, 49, 60, 94
bark 13, 33, 69, 76, 94
biennials 32, 36, 82, 94
branches 9, 10, 12–13, 18, 76–8
buds 13, 14, 22, 30–1, 32, 36, 57, 58–9, 68, 69, 94
bulbs 30–1, 94

cambium 76
carbon dioxide 16, 47, 49
cells 8–16, 18, 20–2, 24, 26, 28, 32–5, 38, 40–1, 43, 44–9, 52, 53, 57, 65
cellulose 8, 45, 95
cereals 68, 70, 72–3, 83, 85
chlorophyll 8, 14–16, 28, 33, 47, 48, 94
chloroplasts 16, 32–3
chromosomes 32–3
climbers 78–9, 91, 94
coal 6, 16, 18
colonizers 56, 59
cones 22–3, 76, 80–1, 94
conservation 92–3
corms 30–1, 94
crops 68, 70, 72–3, 78, 83, 84–6, 92
cytoplasm 8, 18

diseases 30–1
drugs from plants 74–5, 82

ecosystems 92
embryo plant 26–8
energy 6, 16, 18
epidermis 14
evolution 6, 36, 92–3

ferns 6–7, 14, 20–1, 37, 54–5, 57, 58, 94
fertilization 24, 43, 49, 57
fertilizers 43, 56
flagella 39, 47
flowering plants 6, 8, 9, 12–13, 20, 22–6, 30–1, 34, 37, 58–9, 60–6, 69, 72, 76–8, 80, 88, 91, 94, 95
foodstuffs 8–18, 20, 26, 28, 30, 33, 39, 42–3, 45–6, 48, 49, 52, 58–60, 72–3, 83, 84, 95
fossils 6–7, 16, 42, 45
fronds 54–5
fruit 26–8, 30, 35, 94
fungi 6, 11, 16, 17, 37, 38, 42, 48–51, 52–3, 54, 75, 80, 85, 94, 95

gametes 20
gardening 90–1
geotropism 35

germination 20, 28–9, 32, 33, 34, 57, 88–9, 93
glucose 16
grasses 63, 66, 68–9, 72–3, 94
grassland 70
growth 10–35, 39, 48, 76

herbaceous plants 80–4, 94, 95
herbs 82–3, 94
hormones 32, 35
horsetails 6, 54, 95
hybridization 76, 84–5, 95
hydrogen 16
hydroponics 91
hyphae 48, 50, 51, 53, 95

insecticides 74

kelp 43

leaves 6, 8, 9, 10, 12–14, 16–22, 28, 30–2, 35, 43, 48, 56–9, 64–6, 69, 72, 94, 95
lichens 52–3, 63, 80, 95
life, length of plants' 33–4
light 14, 16, 34–5, 42, 54, 62, 80, 93
lignin 9
liverworts 6, 20, 37, 47, 56–7, 58, 80, 95

meristems 10, 21, 95
minerals 9, 10–11, 12, 26, 28, 45, 52, 53, 56, 60–1, 63, 70, 86
mosses 6, 20, 37, 47, 54, 56–7, 58, 60, 63, 80, 95

nectar 25, 95
nitrates 39, 60
nitrogen 39, 60
nodes 12, 95
nodules 39

oil 45
organ systems 9
organisms 6, 8
oxygen 6, 16–18, 28, 39, 58, 60

palms 29, 69, 95
parasites 39–41, 78, 95
peat 16, 18, 56, 60
penicillin 38, 48, 49
perennials 32–3, 36, 66, 76–9, 82, 95
petals 22, 25, 66, 68, 95
photosynthesis 16–18, 28, 39, 47, 48, 95
phototropism 35
plantlets 20–1
plants
  behaviour of 34–5
  classification of 36–7
  evolution of 6–7, 36, 92–3
  fossilized 6–7, 16, 18, 42, 45
  growing of 88–91
  origins of 6–7
  seaside 62–3
  structure of 8–9

upland 62–3
water 58–61, 66–7
pollen 22, 24–6, 58, 72, 76, 94, 95
pollination 22, 24–5, 58, 84–5
pores 13, 14, 16, 18–19, 33, 64

reproduction 6, 9, 13–16, 20–9, 30, 34, 37, 40, 43, 47, 49, 53, 57, 58
rhizomes 13, 55, 95
roots 8–12, 18–21, 26, 28, 30, 32, 35, 39, 43, 48, 52, 56–7, 59, 60–1, 62, 63, 64, 68, 70, 86, 95
runners 13, 33

sap 76, 95
seeds 20, 22–4, 26–9, 32–3, 34, 35, 70, 72, 76, 84, 93, 94, 95
sepals 22, 66, 95
shoots 20–1, 28, 30, 32, 95
shrubs 9, 12–13, 32, 36, 70, 75, 78–80
spikelets 68
spores 20–1, 28, 38, 39, 48–51, 53, 54, 57, 95
stalks, male and female 22, 24–6, 72
stamens 9, 22
stems 9–14, 18–19, 21, 22, 29, 30–1, 33, 35, 48, 56–7, 58, 59, 61, 64–5, 66, 68, 69, 72, 76, 94, 95
stigmas 22, 24–5, 58, 95
stipule 14
stomata 14, 18, 95
succulents 12, 64–5, 94, 95
sugar 16, 18, 26, 28, 30, 34, 68
symbiosis 52

tannins 14
temperature 34, 38–9, 49, 62, 64, 66, 70, 72–3, 77, 80, 86, 93
tendrils 13, 14, 78–9, 95
testa 26, 28
traumatin 35
trees 6–7, 9–15, 18, 24–5, 30, 32, 36, 37, 46, 52–3, 61, 62–3, 69, 70, 76–8, 80–1
tropism 35
trunks 9, 76–9
tubers 30–1, 95

veins 9–15, 18–19, 21, 33, 43, 54, 56, 76
virus 40–1

water 9–19, 25, 26, 28, 30, 32, 33, 34–5, 38, 39, 47, 52, 56, 58–61, 63, 64, 65–6, 72–3, 77, 86, 90, 95
weeds 86–7
wind 24–5, 52, 62, 70, 77, 78

Acknowledgements

The Publishers gratefully acknowledge permission to reproduce the following illustrations:

A-Z Botanical 23b, 45c, r, 52tl, 60, 61tl, bl; Aerofilms Limited 81b; Heather Angel 17bc, 20, 24br, 27b, 29bl, 34, 35bl, br, 49b, 56, 57, 74; Ardea 17br, 29bc, 35tl, tr, 42, 46, 49tl, tr, 52tr, 59tc, tr, bc, br, 63, 66l, 67, 75c, 87t; Dr Lloyd M Beidler/Science Photo Library 8, 18l, 24tl; C M Clay/National Vegetable Research Station 41; Bruce Coleman Limited 6, 7, 25, 27t, 29t, 31, 44, 47r, 51, 55tr, br, 58, 59tl, 62, 64, 66r, 68, 71tl, tr, br, 75r, 82, 83, 87bl, br, 91, 93bl, bl, br; Gene Cox 9bl, br, 12, 15cr, bl, 28, 32, 33, 39, 45l, 47l, 48b; Gene Cox/University of Bath 38; Aubrey Dewar 3, 17t, 23tc, tr; Eastman Kodak Company and Kodak Limited 23tl; Dr Patrick Echlin/University of Cambridge 24tr, bl; Brian Furner 30, 85, 88; N Gryspeerdt 89tl; Peter Hunt 31b, 61r, 69l, 93tr; R D Hunt 15cl, 27c, 75l; Alan Hutchison Library 10, 54; Long Ashton Research Station/Science Photo Library 9tl, tc, tr, 18r, 69r; Massey Ferguson 72–73; Oxford Scientific Films Limited 55l; Dr A W Robards/Department of Biology, University of York 11b; Syndication International 89tr; John Topham Picture Library 11t, 13, 17bl, 21bl, bc, br, 29br, 35cl, cr, 48t, 52b, 71bl, 81t, 90; Trewin Copplestone Publishing Limited 89br; H H Heunert/Zeiss Information 21t.

Jacket photographs: front, Bruce Coleman; back, A-Z Botanical.

Artwork by: Linda Broad 33, 40, 43, 51, 53, 64–65, 68, 73; Carol Kane 8, 16, 37, 41, 77, 79b, 80, 84; David and Theo Nockels 11, 13, 22, 26, 31; Paddy Sellars 14–15, 19, 38, 50, 78–79.